T0331890

# Mathematics of Harmony as a New Interdisciplinary Direction and "Golden" Paradigm of Modern Science

## Volume 1

### The Golden Section, Fibonacci Numbers, Pascal Triangle, and Platonic Solids

K&E Series on Knots and Everything — Vol. 65

# Mathematics of Harmony as a New Interdisciplinary Direction and "Golden" Paradigm of Modern Science

## Volume 1
### The Golden Section, Fibonacci Numbers, Pascal Triangle, and Platonic Solids

## Alexey Stakhov

International Club of the Golden Section, Canada & Academy of Trinitarism, Russia

**World Scientific**

EW JERSEY · LONDON · SINGAPORE · BEIJING · SHANGHAI · HONG KONG · TAIPEI · CHENNAI · TOKYO

*Published by*

World Scientific Publishing Co. Pte. Ltd.
5 Toh Tuck Link, Singapore 596224
*USA office:* 27 Warren Street, Suite 401-402, Hackensack, NJ 07601
*UK office:* 57 Shelton Street, Covent Garden, London WC2H 9HE

Library of Congress Control Number: 2020010957

**British Library Cataloguing-in-Publication Data**
A catalogue record for this book is available from the British Library.

Series on Knots and Everything — Vol. 65
MATHEMATICS OF HARMONY AS A NEW INTERDISCIPLINARY DIRECTION
AND "GOLDEN" PARADIGM OF MODERN SCIENCE
Volume 1: The Golden Section, Fibonacci Numbers, Pascal Triangle, and Platonic Solids

ISBN  978-981-120-710-5 (hardcover)
ISBN  978-981-120-637-5 (ebook for institutions)
ISBN  978-981-120-638-2 (ebook for individuals)

For any available supplementary material, please visit
https://www.worldscientific.com/worldscibooks/10.1142/11445#t=suppl

Desk Editor: Liu Yumeng

Typeset by Stallion Press
Email: enquiries@stallionpress.com

Printed in Singapore

*In fond memory of Yuri Alekseevich Mitropolskiy
and
Alexander Andreevich Volkov*

# Contents

# Preface to the Three-Volume Book

## Continuity in the Development of Science

Scientific and technological progress has a long history and passed in its historical development several stages: The Babylonian and Ancient Egyptian culture, the culture of Ancient China and Ancient India, the Ancient Greek culture, the Middle Ages, the Renaissance, the Industrial Revolution of the 18th century, the Great Scientific Discoveries of the 19th century, the Scientific and Technological Revolution of the 20th century and finally the 21st century, which opens a new era in the history of mankind, the *Era of Harmony*.

Although each of the mentioned stages has its own specifics, at the same time, every stage necessarily includes the content of the preceding stages. This is called the *continuity* in the development of science.

It was during the ancient period, a number of the fundamental discoveries in mathematics were made. They exerted a determining influence on the development of the material and spiritual culture. We do not always realize their importance in the development of mathematics, science, and education. To the category of such discoveries, first of all, we must attribute the *Babylonian numeral system with the base* 60 and the *Babylonian positional principle of number representation,* which is the foundation of the, *decimal, binary, ternary,* and other positional numeral systems. We must add to this list the *trigonometry* and the *Euclidean geometry,* the *incommensurable segments* and the *theory of irrationality,* the *golden section* and *Platonic solids*, the *elementary number theory* and the *mathematical theory of measurement*, and so on.

The *continuity* can be realized in various forms. One of the essential forms of its expression are the fundamental scientific ideas, which permeate all stages of the scientific and technological progress and influence various areas of science, art, philosophy, and technology. The idea of *Harmony*, connected with the *golden section*, belongs to the category of such fundamental ideas.

According to B.G. Kuznetsov, the researcher of Albert Einstein's creativity, the great physicist piously believed that science, physics in particular, always had its eternal fundamental goal *"to find in the labyrinth of the observed facts the objective harmony"*. The deep faith of the outstanding physicist in the existence of the universal laws of the *Harmony* is evidenced by another well-known Einstein's statement: *"The religiousness of the scientist consists in the enthusiastic admiration for the laws of the Harmony"* (the quote is taken from the book *Meta-language of Living Nature* [1], written by the outstanding Russian architect Joseph Shevelev, known for his research in the field of *Harmony* and the *golden section* [1–3]).

## Pythagoreanism and Pythagorean MATHEM's

By studying the sources of the origin of mathematics, we inevitably come to Pythagoras and his doctrine, named the *Pythagoreanism* (see Wikipedia article *Pythagoreanism*, the Free Encyclopedia). As mentioned in Wikipedia, the *Pythagoreanism* originated in the 6th century BC and was based on teachings and beliefs of Pythagoras and his followers called the Pythagoreans. Pythagoras established the first Pythagorean community in Croton, Italy. The Early Pythagoreans espoused a rigorous life and strict rules on diet, clothing and behavior.

According to tradition, *Pythagoreans* were divided into two separate schools of thought: the *mathematikoi* (*mathematicians*) and the *akousmatikoi* (*listeners*). The *listeners* had developed the religious and ritual aspects of *Pythagoreanism*; the *mathematicians* studied the four Pythagorean MATHEMs: *arithmetic*, *geometry*, *spherics*, and *harmonics*. These MATHEMs, according to Pythagoras, were the main composite parts of mathematics. Unfortunately, the Pythagorean MATHEM of the *harmonics* was lost in mathematics during the process of its historical development.

## Proclus Hypothesis

The Greek philosopher and mathematician Proclus Diadoch (412–485 AD) put forth the following unusual hypothesis concerning Euclid's *Elements*. Among Proclus's mathematical works, his *Commentary on the Book I of Euclid's Elements* was the most well known. In the commentary, he puts forth the following unusual hypothesis.

It is well known that Euclid's *Elements* consists of 13 books. In those, XIII$^{th}$ book, that is, the concluding book of the *Elements*, was devoted to the description of the geometric theory of the five *regular polyhedra*, which had played a dominant role in *Plato's cosmology* and is known in modern science under the name of the *Platonic solids*.

Proclus drew special attention to the fact that the concluding book of the *Elements* had been devoted to the *Platonic solids*. Usually, the most important material, of the scientific work is placed in its final part. Therefore, by placing *Platonic solids* in Book XIII, that is, in the concluding book of his *Elements*, Euclid clearly pointed out on main purpose of writing his *Elements*. As the prominent Belarusian philosopher Edward Soroko points out in [4], according to Proclus, Euclid *"had created his Elements allegedly not for the purpose of describing geometry as such, but with purpose to give the complete systematized theory of constructing the five Platonic solids; in the same time Euclid described here some latest achievements of mathematics"*.

It is for the solution of this problem (first of all, for the creation of geometric theory of *dodecahedron*), Euclid already in Book II introduces Proposition II.11, where he describes the *task of dividing the segment in the extreme and mean ratio* (the *golden section*), which then occurs in other books of the *Elements*, in particular in the concluding book (XIII Book).

But the *Platonic solids* in *Plato's cosmology* expressed the *Universal Harmony* which was the main goal of the ancient Greeks science. With such consideration of the *Proclus hypothesis*, we come to the surprising conclusion, which is unexpected for many historians of mathematics. According to the *Proclus hypothesis*, it turns out that from Euclid's *Elements*, two branches of mathematical sciences had originated: the **Classical Mathematics**, which included the *Elements* of the *axiomatic approach* (Euclidean axioms), *the elementary number theory*, and *the theory of irrationalities*, and the Mathematics of Harmony, which was based on the geometric *"task of dividing the segment in the extreme and mean ratio"* (the *golden section*) and also on the theory of the *Platonic solids*, described by Euclid in the concluding Book XIII of his *Elements*.

## The Statements by Alexey Losev and Johannes Kepler

What was the main idea behind ancient Greek science? Most researchers give the following answer to this question: **The idea of Harmony connected to the *golden section*.** As it is known, in ancient Greek philosophy, *Harmony* was in opposition to the *Chaos* and meant the organization of the Universe, the Cosmos. The outstanding Russian philosopher Alexey Losev (1893–1988), the researcher in the aesthetics of the antiquity and the Renaissance, assesses the main achievements of the ancient Greeks in this field as follows [5]:

> *"From Plato's point of view, and in general in the terms of the entire ancient cosmology, the Universe was determined as the certain proportional whole, which obeys to the law of the harmonic division, the golden section ... The ancient Greek system of the cosmic proportion in the literature is often interpreted as the curious result of the unrestrained and wild imagination. In such explanation we see the scientific helplessness of those, who claim this. However, we can understand this historical and aesthetic phenomenon only in the connection with the holistic understanding of history, that is, by using the dialectical view on the culture and by searching for the answer in the peculiarities of the ancient social life."*

Here, Losev formulates the *"golden" paradigm* of ancient cosmology. This paradigm was based upon the fundamental ideas of ancient science that are sometimes treated in modern science as the *"curious result of the unrestrained and wild imagination"*. First of all, we are talking about the *Pythagorean Doctrine of the Numerical Universal Harmony* and *Plato's Cosmology* based on the *Platonic solids*. By referring to the geometrical structure of the Cosmos and its mathematical relations, which express the Cosmic Harmony, the Pythagoreans had anticipated the modern mathematical basis of the natural sciences, which began to develop rapidly in the 20th century. Pythagoras's and Plato's ideas about the Cosmic Harmony proved to be immortal.

Thus, the idea of Harmony, which underlies the ancient Greek doctrine of Nature, was the main "paradigm" of the Greek science, starting from Pythagoras and ending with Euclid. This paradigm relates directly to the *golden section* and the *Platonic solids*, which are the most important Greek geometric discoveries for the expression of the Universal Harmony.

Johannes Kepler (1571–1630), the prominent astronomer and the author of "Kepler's laws", expressed his admiration with the *golden ratio* in the following words [6]:

> *"Geometry has the two great treasures: the first of them is the theorem of Pythagoras; the second one is the division of the line in the extreme and mean ratio. The first one we may compare to the measure of the gold; the second one we may name the precious stone."*

We should recall again that the ancient *task of dividing line segment in extreme and mean ratio* is Euclidean language for the *golden section*!

The enormous interest in this problem in modern science is confirmed by the rather impressive and far from the complete list of books and articles on this subject, published in the second half of the 20th century and the beginning of the 21st century [1–100].

## Ancient Greeks Mathematical Doctrine of Nature

According to the outstanding American historian of mathematics, Morris Kline [101], the main contribution of the ancient Greeks is the one *"which had the decisive influence on the entire subsequent culture, was that they took up the study of the laws of Nature".* The main conclusion, from Morris Kline's book [101] is the fact that the ancient Greeks proposed the innovative concept of the Cosmos, in which everything was subordinated to the mathematical laws. Then the following question arises: during which time this concept was developed? The answer to this question is also addressed in Ref. [101].

According to Kline [101], the innovative concept of the Cosmos based on the mathematical laws, was developed by the ancient Greeks in the period from VI to III centuries BC. But according to the prominent Russian mathematician academician A.N. Kolmogorov [102], in the same period in ancient Greece, *"the mathematics was created as the independent science with the clear understanding of the uniqueness of its method and with the need for the systematic presentation of its basic concepts and proposals in the fairly general form."* But then, the following important question, concerning the history of the original mathematics arises: was there any relationship between the process of creating the mathematical theory of Nature, which was considered as the goal and the main achievement of ancient Greek science [101], and the process of creating mathematics, which happened in ancient Greece in the same period [102]? It turns out that such connection, of course, existed. Furthermore, it can be argued that these processes actually coincided, that is, the processes of the creation of mathematics by the ancient Greeks [102], and their doctrine of Nature, based on the mathematical principles [101], were one and the same processes. And the most vivid embodiment of the process of the *Mathematization of Harmony* [68] happened in Euclid's *Elements*, which was written in the third century BC.

## Introduction of the Term *Mathematics of Harmony*

In the late 20th century, to denote the mathematical doctrine of Nature, created by the ancient Greeks, the term *Mathematics of Harmony* was introduced. It should be noted that this term was chosen very successfully because it reflected the main idea of the ancient Greek science, the *Harmonization of Mathematics* [68]. For the first time, this term was introduced in the small article "Harmony of spheres", placed in *The Oxford Dictionary of Philosophy* [103]. In this article, the concept of *Mathematics of Harmony* was associated with the *Harmony of spheres*, which was, also called in Latin as *"harmonica mundi"* or *"musica mundana"* [10]. The *Harmony of spheres* is the ancient and medieval doctrine on the musical and mathematical structure of the Cosmos, which goes back to the Pythagorean and Platonic philosophical traditions.

Another mention about the *Mathematics of Harmony* in the connection to the ancient Greek mathematics is found in the book by Vladimir Dimitrov, *A New Kind of Social Science*, published in 2005 [44]. It is important to emphasize that in Ref. [44], the concept of *Mathematics of Harmony* is directly associated with the *golden section*, the most important mathematical discovery of the ancient science in the field of Harmony. This discovery at that time was called *"dividing a segment into the extreme and mean ratio"* [32].

From Refs. [44, 45], it is evident that prominent thinkers, scientists and mathematicians took part in the development of the *Mathematics of Harmony* for several millennia: Pythagoras, Plato, Euclid, Fibonacci, Pacioli, Kepler, Cassini, Binet, Lucas, Klein, and in the 20th century the well-known mathematicians Coxeter [7], Vorobyov [8], Hoggatt [9], Vaida [11], Knuth [123], and so on. And we cannot ignore this historical fact.

## Fibonacci Numbers

The Fibonacci numbers, introduced into Western European mathematics in the 13th century by the Italian mathematician Leonardo of Pisa (known by the nickname Fibonacci), are closely related to the *golden ratio*. Fibonacci numbers from the numerical sequence, which starts with two units, and then each subsequent Fibonacci number is the sum of the two previous ones: $1, 1, 2, 3, 5, 8, 13, 21, 34, 55, \ldots$. The ratio of the two neighboring Fibonacci numbers in the limit tends to be the *golden ratio*.

The mathematical theory of Fibonacci numbers has been further developed in the works of the French mathematicians of the 19th century Binet (*Binet formula*) and Lucas (*Lucas numbers*). As mentioned above, in the second half of the 20th century, this theory was developed in the works of the Canadian geometer, Donald Coxeter [7], the Soviet mathematician, Nikolay Vorobyov [8], the American mathematician, Verner Hoggatt [9] and the English mathematician, Stefan Vajda [11], the outstanding American mathematician, Donald Knuth [123], and so on.

The development of this direction ultimately led to the emergence of the *Mathematics of Harmony* [6], a new interdisciplinary direction of modern science that relates to modern mathematics, computer science, economics, as well as to all theoretical natural sciences. The works of the well-known mathematicians, Coxeter [7], Vorobyov [8], Hoggatt [9], Vaida [11], Knuth [123], and others, as well as the study of Fibonacci mathematicians, members of the American Fibonacci Association, became the beginning of the process of *Harmonization of Mathematics* [68], which continues actively in the 21st century. And this process is confirmed by a huge number of books and articles in the field of the *golden section* and *Fibonacci numbers* published in the second half of the 20th century and the beginning of the 21st century [1–100].

### Sources of the Present Three-Volume Book

The differentiation of modern science and its division into separate spheres do not allow us often to see the general picture of science and the main trends in its development. However, in science, there exist research objects that combine disparate scientific facts into a single whole. *Platonic solids* and the *golden section* are attributed to the category of such objects. The ancient Greeks elevated them to the level of *"the main harmonic figures of the Universe"*. For centuries or even millennia, starting from Pythagoras, Plato and Euclid, these geometric objects were the object of admiration and worship of the outstanding minds of mankind, during Renaissance, Leonardo da Vinci, Luca Pacoli, Johannes Kepler, in the 19th century, Zeising, Lucas, Binet and Klein. In the 20th century, the interest in these mathematical objects increased significantly, thanks to the research of the Canadian geometer, Harold Coxeter [7], the Soviet mathematician Nikolay Vorobyov [8] and the American mathematician Verner Hoggatt [9], whose works in the field of the Fibonacci numbers began the process of the "Harmonization of Mathematics". The development of this direction led to the creation of the *Mathematics of Harmony* [6] as a new interdisciplinary trend of modern science.

The newest discoveries in the various fields of modern science, based on the *Platonic solids*, the *golden section* and the *Fibonacci numbers*, and new scientific discoveries and mathematical results, related to the *Mathematics of Harmony* (*quasicrystals* [115], *fullerenes* [116], the new geometric theory of phyllotaxis (*Bodnar's geometry*) [28], the *general theory of the hyperbolic functions* [75, 82], the *algorithmic measurement theory* [16], the *Fibonacci and golden ratio codes* [6], the *"golden" number theory* [94], the *"golden" interpretation of the special theory of relativity* and the *evolution of the Universe* [87], and so on) create an overall picture of the movement of modern science towards the *"golden" scientific revolution*, which is one of the characteristic trends in the development of modern science. The sensational information about the experimental discovery of the golden section in the quantum world as a result of many years of research, carried out at the Helmholtz–Zentrum Berlin für Materialien und Energie (HZB) (Germany), the Oxford and Bristol Universities and the Rutherford Appleton Laboratory (UK), is yet another confirmation of the movement of the theoretical physics to the *golden section* and the *Mathematics of Harmony* [6].

For the first time, this direction was described in the book by Stakhov A.P., assisted by Scott Olsen, *The Mathematics of Harmony. From Euclid to Contemporary Mathematics and Computer Science*, World Scientific, 2009 [6].

In 2006, the Russian Publishing House, "Piter" (St. Petersburg) published the book, *Da Vinci Code and Fibonacci numbers* [46] (Alexey Stakhov, Anna Sluchenkova and Igor Shcherbakov were the authors of the book). This book was one of the first Russian books in this field. Some aspects of this direction are reflected in the following authors' books, published by Lambert Academic Publishing (Germany) and World Scientific (Singapore):

- Alexey Stakhov, Samuil Aranson, *The Mathematics of Harmony and Hilbert's Fourth Problem. The Way to Harmonic Hyperbolic and Spherical Worlds of Nature.* Germany: Lambert Academic Publishing, 2014 [51].
- Alexey Stakhov, Samuil Aranson, Assisted by Scott Olsen, *The "Golden" Non-Euclidean Geometry*, World Scientific, 2016 [52].
- Alexey Stakhov, *Numeral Systems with Irrational Bases for Mission-Critical Applications*, World Scientific, 2017 [53].

These books are fundamental in the sense that they are the first books in modern science, devoted to the description of the theoretical foundations and applications of the following new trends in modern science: the *history of the golden section* [78], the *Mathematics of Harmony* [6], the *"Golden" Non-Euclidean geometry* [52], ascending to Euclid's *Elements*, and also the *Numeral Systems with Irrational bases*, ascending to the Babylonian positional numeral system, the decimal and binary system and Bergman's system [54].

These books discuss the problems, which in modern mathematics are considered long resolved and therefore are not included in the circle of the studies of mathematicians, namely the new mathematical theory of measurement called the *Algorithmic Measurement Theory* [16, 17], the *Mathematics of Harmony* [6] as a new kind of elementary mathematics that has a direct relationship to the foundations of the mathematics and mathematical education, the new class of the elementary functions called the *hyperbolic Fibonacci and Lucas functions* [64, 75, 82] and finally, the new ways of real numbers representation, and the new binary and ternary arithmetic's [55, 72], which have the fundamental interest for computer science and digital metrology.

In 2010, the Odesa I.I. Mechnikov National University (Ukraine) hosted the *International Congress on the Mathematics of Harmony*. The main goal of the Congress was to consolidate the priority of Slavic science in the development of this important trend and acquaint the scientific community with the main trends of the development of the Mathematics of Harmony as the new interdisciplinary direction of modern science.

In the recent years, the new unique books on the problems of Harmony and the history of the golden section have been published:

- *The Prince of Wales. Harmony. A New Way of Looking at our World* (*coauthors Tony Juniper and Ian Skelly*). An Imprint of HarperCollins Publisher, 2010 [49].
- Hrant Arakelian, *Mathematics and History of the Golden Section.* Moscow, Publishing House "Logos", 2014 [50].

For the last 30 years, *Charles, The Prince of Wales*, had been known around the world as one of the most forceful advocates for the environment. During that period, he focused on many different aspects of our lives, when we continually confront with the real life from new angles of view and search original approaches. Finally, in *Harmony* (2010) [49], The Prince of Wales and his coauthors laid out their thoughts on the planet, by offering an in-depth look into its future. Here, we see a dramatic call to the action and an inspirational guide on the relationship of mankind with Nature throughout history. The Prince of Wales's *Harmony* (2010) [49] is an illuminating look on how we must reconnect with our past in order to take control of our future.

The 2014 book [50] by the Armenian philosopher and physicist *Hrant Arakelian* is devoted to the *golden section* and to the complexity of problems connected with it. The book consists of two parts. The first part is devoted to the mathematics of the golden section and the second part to the history of the golden section. Undoubtedly, Arakelian's 2014 book is one of the best modern books devoted to mathematics and the history of the golden section.

The *International Congress on Mathematics of Harmony* (Odessa, 2010) and the above-mentioned books by *The Prince of Wales* and Armenian philosopher *Hrant Arakelian* are brilliant confirmation of the fact that in modern science, the interest in the mathematics of the golden section and its history increases and further development of the Mathematics of Harmony can lead to revolutionary transformations in modern mathematics and science on the whole.

Why did the author decide to write the three-volume book *The Mathematics of Harmony as a New Interdisciplinary Direction and "Golden" Paradigm of Modern Science*? It should be noted that the author and other famous authors in this field published many original books and articles in this scientific direction. However, all the new results and ideas, described in the above-mentioned publications of Alexey Stakhov, Samuil Aranson, Charles, The Prince of Wales, Hrant Arakelian and other authors are scattered in their numerous articles and books, which makes it difficult to understand their fundamental role in the development of the modern mathematics, computer science and theoretical natural sciences on the whole.

This role is most clearly reflected in the following citations taken from *Harmony* by the Prince of Wales (2010) [49]:

> *"This is a call to revolution. The Earth is under threat. It cannot cope with all that we demand of it. It is losing its balance and we humans are causing this to happen."*

The following quote, placed on the back cover of Prince of Wales's *Harmony* [49], develops this thought:

> *"We stand at an historical moment; we face a future where there is a real prospect that if we fail the Earth, we fail humanity. To avoid such an outcome, which will comprehensively destroy our children's future or even our own, we must make choices now that carry monumental implications."*

Thus, *The Prince of Wales* has considered his 2010 book, *Harmony. A New Way of Looking at our World*, as a call to the revolution in modern science, culture and education. The same point of view is expressed in the above-mentioned books by Stakhov and Aranson [6, 46, 51–53]. Comparing the books of *Prince of Wales* [49] and *Hrant Arakelian* [50] to the 2009, 2016 and 2017 books of Alexey Stakhov and Samuil Aranson [6, 51–53], one can only be surprised how deeply all these books, written in different countries and continents, coincide in their ideas and goals.

Such an amazing coincidence can only be explained by the fact that in modern science, there is an urgent need to return to the "harmonious ideas" of Pythagoras, Plato and Euclid that permeated across the ancient Greek science and culture. The *Harmony* idea, formulated in the works of the Greek scholars and reflected in Euclid's *Elements* turned out to be immortal!

We can safely say that the above-mentioned books by Stakhov and Aranson (2009, 2016, 2017) [6, 51–53], the book by The Prince of Wales with the coauthors (2010) [49] and book by Arakelian (2014) [50] are the beginning of a revolution in modern science. The essence of this revolution consists, in turning to the fundamental ancient Greek idea of the *Universal Harmony*, which can save our Earth and humanity from the approaching threat of the destruction of all mankind.

It was this circumstance that led the author to the idea of writing the three-volume book *Mathematics of Harmony as a New Interdisciplinary Direction and "Golden" Paradigm of Modern Science*, in which the most significant and fundamental scientific results and ideas, formulated by the author and other authors (The Prince of Wales, Hrant Arakelian, Samuil Aranson and others) in the process of the development of this scientific direction, will be presented in a popular form, accessible to students of universities and colleges and teachers of mathematics, computer science, theoretical physics and other scientific disciplines.

## Structure and the Main Goal of the Three-Volume Book

The book consists of three volumes:

- *Volume I. The Golden Section, Fibonacci Numbers, Pascal Triangle and Platonic Solids.*
- *Volume II. Algorithmic Measurement Theory, Fibonacci and Golden Arithmetic and Ternary Mirror-Symmetrical Arithmetic.*
- *Volume III. The "Golden" Paradigm of Modern Science: Prerequisite for the "Golden" Revolution in the Mathematics, the Computer Science, and Theoretical Natural Sciences.*

Because the *Mathematics of Harmony* goes back to the "harmonic ideas" of Pythagoras, Plato and Euclid, the publication of such a three-volume book will promote the introduction of these "harmonic ideas" into modern education, which is important for more in-depth understanding of the ancient conception of the *Universal Harmony* (as the main conception of ancient Greek science) and its effective applications in modern mathematics, science and education.

The main goal of the book is to draw the attention of the broad scientific community and pedagogical circles to the *Mathematics of Harmony*, which is a new kind of elementary mathematics and goes back to Euclid's *Elements*. The book is of interest for the modern mathematical education and can be considered as the "golden" paradigm of modern science on the whole.

The book is written in a popular form and is intended for a wide range of readers, including schoolchildren, school teachers, students of colleges and universities and their teachers, and also scientists of various specializations, who are interested in the history of mathematics, Platonic solids, golden section, Fibonacci numbers and their applications in modern science.

# Introduction

It is known that the amount of irrational (incommensurable) numbers is infinite. However, some of them occupy a special place in the history of mathematics, science and education. Their significance lies in the fact that they are expressing some fundamental relationships, which are universal by their nature and appear in the most unexpected places.

The first of them is the irrational number $\sqrt{2}$ equal to the ratio of the diagonal to the side of a square. This number is associated with the discovery of "incommensurable segments" and the history of the most dramatic period in ancient mathematics that led to the development of the *theory of irrationalities* and *irrational numbers* and, ultimately, to the creation of modern "continuous" mathematics.

The next two irrational (transcendental) numbers are as follows: the number of $\pi$, which is equal to the ratio of the length of circumference to its diameter (this number lies at the basis of the *trigonometric functions*) and the *Naperian number* of $e$ (this number underlies the *hyperbolic functions* and is the basis of *natural logarithms*). Between $\pi$ and $e$, that is, between the two irrational numbers that dominate over the analysis, there is the following elegant relation derived by Euler:

$$1 + e^{i\pi} = 0,$$

where $i = \sqrt{-1}$ is an *imaginary unit*, another amazing creation of the mathematical mind.

Another famous irrational number is the *"golden proportion"* $\Phi = (1 + \sqrt{5})/2$, which arises as a result of solving the geometric task of *"dividing a segment in the extreme and mean ratio"* [32]. This task is described in Book II of Euclid's *Elements* (Proposition II.11).

In the preface, we already mentioned about the brilliant German astronomer, Johannes Kepler, who named the *golden ratio* as one of the "treasures of geometry" and compared it to the *Pythagoras theorem.* A prominent Soviet philosopher Alexey Losev, a researcher of the aesthetics of antiquity and the Renaissance, in his citation (see the preface) argues that *"from Plato's point of view, and in general in terms of the entire ancient cosmology, the Universe is determined as a certain proportional whole, which obeys to the law of harmonic division, the golden section."*

It is well known that the Fibonacci numbers $1, 1, 2, 3, 5, 8, 13, \ldots$, introduced in the 13th century by the famous Italian mathematician Leonardo of Pisa (Fibonacci) for solving the "task of rabbits' reproduction" are closely related to the *golden ratio.* The deep mathematical connection between the *Fibonacci numbers* and the *golden ratio* is that the ratio of the two neighboring Fibonacci numbers in the limit tends to be the *golden ratio*, which implies that this numerical sequence is also expressing *Harmony.*

The so-called *Pascal's triangle*, the special table for the location of binomial coefficients, is one of the highly harmonious objects of mathematics. This table was proposed in the 17th century by the outstanding French mathematician and physicist Blaise Pascal (1623–1662). In the second half of the 20th century, the famous American mathematician George Polya (1887–1995) in his book [111] had described the connection of *Fibonacci numbers* to the so-called *diagonal sums of Pascal's triangle.* The development of these ideas led to a generalization of the *task of rabbit reproduction* and the introduction of the so-called *Fibonacci p-numbers* [6].

Volume I of this three-volume book is devoted to the presentation of the foundations of the theory of these extremely beautiful mathematical objects (*Fibonacci p-numbers*), the interest in which will not fade for centuries or even millennia.

Volume I consists of four chapters. Chapter 1 "The Golden Section: History and Applications" begins with the analysis of a sensational hypothesis, the *Proclus hypothesis*, which overturns our ideas about Euclid's *Elements* and the entire history of origin of the mathematics. According to this hypothesis, Euclid's *Elements* were created under the powerful influence of the "Harmony idea", which was the basic concept of ancient Greek science. The main goal of *Elements* was to create a complete geometric theory of *Platonic solids*. This theory was described by Euclid in the final (Book XIII) book. To give the completed theory of *dodecahedron*, Euclid, already in Book II, introduces and solves the task of "dividing the segment in the extreme and mean ratio" (Proposition II.11) [32], which in modern science is called the *golden section*.

Chapter 1 deals with the following: the geometric method of constructing the *golden section*, the algebraic equation of the *golden section*, the most famous algebraic identities for the *golden ratio* and also the geometric figures associated with the *golden section* ( *"golden" triangles, pentagon* and *pentagram, "golden" ellipse, decagon*, etc.). Chapter 1 ends with examples of using the *golden section* in works of fine arts and culture ( *Cheops pyramid, the arts of ancient Greece and the Renaissance*).

Chapter 2 is a popular introduction to the "theory of Fibonacci and Lucas numbers", which actively began to develop in the second half of the 20th century in the works of Soviet and Western mathematicians and philosophers [7–11]. In Chapter 2, we expound the little-known results and applications of the Fibonacci and Lucas numbers, such as *Steinhaus's Iron Table*, the connection of Fibonacci numbers with *Pythagorean triangles, numerological properties of Fibonacci and Lucas numbers* and we also consider the examples of applications of Fibonacci numbers in Nature (*pentagonal symmetry, Fibonacci spirals, phyllotaxis phenomenon*, and so on).

In Chapter 2, we also describe the original *theory of Fibonacci p-numbers*, introduced by Alexey Stakhov in the middle of 1960s. The foundations of this theory were expounded by Stakhov in Refs. [6, 16, 17]. Also, the relationship of the Fibonacci *p*-numbers to the Pascal triangle and binomial coefficients is shown.

Chapter 3 is devoted to the discussion of diagonal sums of *Pascal's triangle* to *Fibonacci p-numbers*. In Chapter 3, we consider a *generalization of the problem of the golden section* [6, 16, 17, 60] and introduce an important concept of the *"golden p-proportion"*, which is a positive root of the algebraic equation of the golden p-proportion and generalization of the classical golden proportion. We consider the algebraic equations for the golden p-proportion, based on Vieta's formulas, and also *Binet's formulas for the Fibonacci p-numbers and for the Lucas p-numbers*.

Chapter 4 is devoted to the Platonic solids and Plato's cosmology and to the discussion of the historical role of the Platonic solids in the two outstanding discoveries of modern theoretical natural sciences, *fullerenes* and *quasicrystals*, which were awarded the Nobel Prize. We consider the *Archimedean truncated icosahedron* as the most important geometric model of the *fullerenes*, as the mystery of the Egyptian calendar and its connection to the *dodecahedron* and also Klein's conception of *icosahedron* as the main geometric object of mathematics [113]. In concluding part of Chapter 4, we consider the new ideas in the theory of elementary particles, based on the Platonic solids.

# About the Author

**Alexey Stakhov**, born in May 7, 1939, is a Ukrainian mathematician, inventor and engineer, who has made a contribution to the theory of Fibonacci numbers and the *golden section* and their applications in computer science and measurement theory. He is a Doctor of Computer Science (1972) and a Professor (1974), and the author of over 500 publications, 14 books and 65 international patents. He is also the author of many original publications in computer science and mathematics, including *algorithmic measurement theory* [16, 17], *Fibonacci codes and codes of the golden proportions* [19], *hyperbolic Fibonacci and Lucas functions* [64, 75] and finally the *Mathematics of Harmony* [6], which goes back in its origins to Euclid's *Elements*. In these areas, Alexey Stakhov has written many papers and books, which have been published in famous scientific journals by prestigious international publishers.

The making of Alexey Stakhov as a scientist is inextricably linked with the Kharkov Institute for Radio Electronics, where he was a postgraduate student of the Technical Cybernetics Department from 1963 to 1966. Here, he defended his PhD thesis in the field of Technical Cybernetics (1966) under the leadership of the prominent Ukrainian scientist Professor Alexander Volkov. In 1972, Stakhov defended (at the age of 32 years) his Grand Doctoral dissertation *Synthesis of Optimal Algorithms for Analog-to-Digital Conversion* (Computer Science speciality). Although the dissertation had an engineering character, Stakhov in his books and articles touched upon two fundamental problems of mathematics: *theory of measurement* and *numeral systems*.

Prof. Stakhov worked as "Visiting Professor" of different Universities: Vienna Technical University (Austria, 1976), University of Jena (Germany, 1986), Dresden Technical University (Germany, 1988), Al Fateh University (Tripoli, Libya, 1995–1997), Eduardo Mondlane University (Maputo, Mozambique, 1998–2000).

## Stakhov's Prizes and Awards

- Award for the best scientific publication by Ministry of Education and Science of Ukraine (1980);
- Barkhausen's Commemorative Medal issued by the Dresden Technical University as "Visiting Professor" of Heinrich Barkhausen's Department (1988);
- Emeritus Professor of Taganrog University of Radio Engineering (2004);
- The honorary title of "Knight of Arts and Sciences" (Russian Academy of Natural Sciences, 2009);
- The honorary title "Doctor of the Sacred Geometry in Mathematics" (American Society of the Golden Section, 2010);
- Awarded "Leonardo Fibonacci Commemorative Medal" (Interdisciplinary Journal "De Lapide Philosophorum", 2015).

# Acknowledgments

Alexey Stakhov expresses great thanks to his teacher, the outstanding Ukrainian scientist, Professor Alexander Volkov; under his scientific leadership, the author defended PhD dissertation (1966) and then DSc dissertation (1972). These dissertations were the first steps in Stakhov's research, which led him to the conceptions of *Mathematics of Harmony* and *Fibonacci computers* based on the *golden section* and *Fibonacci numbers*.

During his stormy scientific life, Stakhov met many fine people, who could understand and evaluate his enthusiasm and appreciate his scientific direction. About 50 years ago, Alexey Stakhov had read the remarkable brochure *Fibonacci Numbers* [8] written by the famous Soviet mathematician Nikolay Vorobyov. This brochure was the first mathematical work on, Fibonacci numbers published in the second half of the 20th century. This brochure, determined Stakhov's scientific interest in the *Fibonacci numbers* and the *golden section* for the rest of his life. In 1974, Professor Stakhov met with Professor Vorobyov in Leningrad (now St. Petersburg) and told Professor Vorobyov about his scientific achievements in this area. Professor Vorobyov gave Professor Stakhov, his brochure *Fibonacci Numbers* [8] as a keepsake with the following inscription: "*To highly respected Alexey Stakhov with Fibonacci's greetings*". This brief inscription because a certain kind of guiding star for Alexey Stakhov.

With deep gratitude, Stakhov recollects the meeting with the famous Austrian mathematician Professor *Alexander Aigner* in the Austrian city of Graz in 1976. The meeting with Professor Aigner was the beginning of the international recognition of Stakhov's scientific direction.

Another remarkable scientist, who had a great influence on Stakhov's research, was the Ukrainian mathematician and academician *Yuri Mitropolskiy*, the *Head of the Ukrainian Mathematical School* and the *Chief Editor of the Ukrainian Mathematical Journal*. His influence on Stakhov's researches, pertinent to the history of mathematics and other topics, such as the application of the Mathematics of Harmony in contemporary mathematics, computer science and mathematical education, were inestimable stimulus for Alexey Stakhov. Thanks to the support of Yuri Mitropolskiy, Stakhov published many important articles in the prestigious Ukrainian academic journals, including the *Ukrainian Mathematical Journal*.

In 2002, *The Computer Journal* (British Computer Society) published the fundamental article by Stakhov, *"Brousentsov's Ternary Principle, Bergman's Number System and Ternary Mirror-Symmetrical Arithmetic"* (*The Computer Journal*, Vol. 45, No. 2, 2002) [72]. This article by Stakhov created great interest among all the English scientific computer community. Emeritus Professor of Stanford University *Donald Knuth* was the first outstanding world scientist, who congratulated Prof. Stakhov with this publication. Donald Knuth's letter became one of the main stimulating factors for writing Stakhov's future book *The Mathematics of Harmony From Euclid to Contemporary Mathematics and Computer Science* (World Scientific, 2009) [6]. *Professor Stakhov considers this book as the main book of his scientific life.*

Stakhov's arrival to Canada in 2004 became the beginning of a new stage in his scientific research. Within 15 years (2004–2019), Prof. Stakhov had published more than 50 fundamental articles in different international English-language journals, such as *Chaos, Solitons & Fractals, Applied Mathematics, Arc Combinatoria, The Computer Journal, British Journal of Mathematics and*

*Computer Science, Physical Science International Journal, Visual Mathematics*, etc. Thanks to the support of Prof. El-Nashie, the Editor-in-Chief of the Journal *Chaos, Solitons & Fractals* (UK), Stakhov published in this journal 15 fundamental scientific articles that garnered great interest among the English-speaking scientific community.

The publication of the three fundamental books *The Mathematics of Harmony* (World Scientific, 2009) [6], *The "Golden" Non-Euclidean Geometry* (World Scientific, 2016, co-author Prof. Samuil Aanson) [52] and *Numeral Systems with Irrational Bases for Mission-Critical Applications* (World Scientific, 2017) [53] is one of the main scientific achievements by Stakhov during the Canadian period of his scientific creativity. These books were published thanks to the support of the famous American mathematician Prof. *Louis Kauffman*, Editor-in-Chief of the *Series on Knots and Everything* (World Scientific) and Prof. *M.S. Wong*, the famous Canadian mathematician (York University) and Editor-in-Chief of the *Series on Analysis, Application and Computation* (World Scientific). A huge assistance in the publication of Stakhov's books by of World Scientific was rendered by the American researcher, Scott Olsen, Professor of Philosopy at the College of Central Florida, and Jay Kappraff, Emeritus Professor of Mathematics at the New Jersey Institute of Technology. Prof. Scott Olsen, who was one of the leading US experts in the field of *Harmony* and the *golden section*, was the English editor for Stakhov's book mentioned above and the Emeritus Professor Jay Kappraff was the reviewer of Stakhov's book, *The Mathematics of Harmony* (World Scientific, 2009).

The prominent Ukrainian mathematician and head of the Ukrainian Mathematical School, *Yuri Mitropolskiy*, praised highly Stakhov's *Mathematics of Harmony*. Academician Mitropolsky organized Stakhov's speech at the meeting of the Ukrainian Mathematical Society in 1998. Based upon his recommendation, Stakhov's articles were published in the Ukrainian academic journals, in particular, the *Ukrainian Mathematical Journal*. Under his direct influence, Stakhov started writing the book, *The Mathematics of Harmony. From Euclid to Contemporary Mathematics and Computer*

*Science* [6], which was published by World Scientific in 2009 following the death of the academician Mitropolsky in 2008.

## Scientific cooperation of Alexey Stakhov and Samuil Aranson

Samuil Aranson's acquaintance to the *golden section* and the *Fibonacci numbers* began in 2001 after the reading of a very rare book "Chain Fractions" [107] by the famous Russian mathematician, Aleksandr Khinchin. In this book, Samuil Aranson found results, related to the representation of the *"golden ratio"* in the form of a continued fraction.

In 2007, Prof. Aranson read a wonderful Internet publication, *Museum of Harmony and Golden Section*, posted in 2001 by Professor Alexey Stakhov and his daughter Anna Sluchenkova. This Internet Museum covers various areas of modern natural sciences and tells about the different and latest scientific discoveries, based on the *golden ratio* and *Fibonacci numbers*, including the *Mathematics of Harmony* and its applications in modern natural sciences. After reading this Internet Museum, Samuil Aranson contacted Alexey Stakhov in 2007 through e-mail and offered him joint scientific collaboration in further application of the *Mathematics of Harmony* in various areas of mathematics and modern natural sciences. Scientific collaboration between Alexey Stakhov and Samuil Aranson turned out to be very fruitful and continues up to the present time.

New ideas in the field of elementary mathematics and the history of mathematics, developed by Stakhov (*Proclus's hypothesis* as a new look at Euclid's *Elements* and history of mathematics, *hyperbolic Fibonacci and Lucas functions* [64, 75] as a new class of elementary functions and other mathematical results) attracted the special attention of Prof. Aranson. Scientific collaboration between Stakhov and Aranson began in 2007. From 2007, they published the following joint scientific works (in Russian and English), giving fundamental importance for the development of mathematics and modern theoretical natural sciences:

## Stakhov and Aranson's Mathematical Monographs in English

1. Stakhov A., Aranson S., *The Mathematics of Harmony and Hilbert's Fourth Problem. The Way to the Harmonic Hyperbolic and Spherical Worlds of Nature.* Germany: Lambert Academic Publishing, 2014.
2. Stakhov A., Aranson S., Assisted by Scott Olsen, *The "Golden" Non-Euclidean Geometry: Hilbert's Fourth Problem, "Golden" Dynamical Systems, and the Fine-Structure Constant*, World Scientific, 2016.

## Stakhov and Aranson's Scientific Papers in English

3. Stakhov A.P., Aranson S.Kh., "Golden" Fibonacci goniometry, Fibonacci-Lorentz transformations, and Hilbert's fourth problem. *Congressus Numerantium* **193**, (2008).
4. Stakhov A.P., Aranson S.Kh., Hyperbolic Fibonacci and Lucas functions, "golden" Fibonacci goniometry, Bodnar's geometry, and Hilbert's fourth problem. Part I. Hyperbolic Fibonacci and Lucas functions and "Golden" Fibonacci goniometry. *Applied Mathematics* **2**(1), (2011).
5. Stakhov A.P., Aranson S.Kh., Hyperbolic Fibonacci and Lucas functions, "golden" Fibonacci goniometry, Bodnar's geometry, and Hilbert's fourth problem. Part II. A new geometric theory of phyllotaxis (Bodnar's Geometry). *Applied Mathematics* **2**(2), (2011).
6. Stakhov A.P., Aranson S.Kh., Hyperbolic Fibonacci and Lucas functions, "golden" Fibonacci goniometry, Bodnar's geometry, and Hilbert's fourth problem. Part III. An original solution of Hilbert's fourth problem. *Applied Mathematics* **2**(3), (2011).
7. Stakhov A.P., Aranson S.Kh., The mathematics of harmony, Hilbert's fourth problem and Lobachevski's new geometries for physical world. *Journal of Applied Mathematics and Physics* **2**(7), (2014).

8. Stakhov A., Aranson S., The fine-structure constant as the physical-mathematical millennium problem. *Physical Science International Journal* **9**(1), (2016).

9. Stakhov A., Aranson S., Hilbert's fourth problem as a possible candidate on the millennium problem in geometry. *British Journal of Mathematics & Computer Science* **12**(4), (2016).

# Chapter 1

# The Golden Section: History and Applications

## 1.1. The Idea of the Universal Harmony in Ancient Greek Science

### 1.1.1. What is Harmony?

V.P. Shestakov, the author of the book *Harmony as an Aesthetic Category* [10], notes the following:

> "*In the history of aesthetic teachings, various types of understanding of harmony were put forward. The very concept of "harmony" was used extremely broadly and multivalently. It denoted both the natural structure of Nature and the Cosmos, and the beauty of the physical and moral world of man and the principles of the structure of the work of art, and the laws of aesthetic perception.*"

Shestakov singles out three basic understandings of *Harmony* that evolved in the process of development of science and aesthetics:

(1) **Mathematical understanding of Harmony or mathematical Harmony**. In this sense, harmony is understood as the equality or proportionality of the parts with each other and the part with the whole. In the Great Soviet Encyclopedia, we find the following definition of *Harmony*, which expresses the

mathematical understanding of harmony:

> *"Harmony is the proportionality of parts and the whole, the fusion of various components of an object into a single organic whole. Harmony is the outer revealing of inner order and the measure of existence."*

(2) **Aesthetic harmony**. Unlike the mathematical understanding, the aesthetic understanding is no longer just quantitative, but qualitative, expressing the inner nature of things. The aesthetic *Harmony* is associated with aesthetic experiences, with aesthetic evaluation. This type of harmony is most clearly manifested in the perception of the beauty of Nature.

(3) **Artistic harmony**. This type of *Harmony* is associated with art. The artistic *Harmony* is the actualization of the principle of *Harmony* in the material of art itself.

The most important aspect that follows from the above reasonings is the following: *Harmony is a universal concept that has relation not only to mathematics and science but also to fine arts.*

### 1.1.2. Numerical harmony of Pythagoreans

**Pythagoras** and **Heraclitus** were philosophers and thinkers, whose names are usually associated with the beginning of the philosophical doctrine of *Harmony*. According to many authors, the key idea of *Harmony* as a proportional unity of opposites belongs to Pythagoras. Pythagoreans first put forward the idea of *harmonious construction of the whole world*, including not only Nature and Man but also the entire cosmos. According to Pythagoreans, *"harmony is an inner connection of things, without which the cosmos could not exist"* [10]. Finally, according to Pythagoras, harmony has a numerical expression, that is, it is integrally connected with the concept of number.

**Pythagoras** (570–500 BC) is perhaps one of the most famous scientists in the history of science. He is revered by every person who studies geometry and is familiar with the "Pythagoras theorem", one of the most famous theorems of geometry. In ancient literature,

Fig. 1.1. Pythagoras.

Pythagoras has been described by his contemporaries as a well-known philosopher and scholar, a religious and ethical reformer, an influential politician, a demigod in the eyes of his disciples and a charlatan. His popularity was such that during his lifetime, coins with his image were issued in 430–420 BC. For the fifth century BC, this was an unprecedented case! Pythagoras was the first Greek philosopher awarded to a special assay (Fig. 1.1).

The important role of Pythagoras in the development of Greek science consists in the fact that he fulfilled a historical mission in transferring the knowledge of the Egyptian and Babylonian priests into the culture of Ancient Greece. It was thanks to Pythagoras, who undoubtedly was one of the most educated thinkers of his time, that Greek science received a huge amount of knowledge in the fields of philosophy, mathematics and natural sciences, which, by getting into the favorable environment of ancient Greek culture, contributed to its rapid development.

Pythagoreans created the doctrine of the creative essence of the number. Aristotle in "Metaphysics" notes this particular feature of the Pythagorean doctrine:

> *"The so-called Pythagoreans, having engaged in mathematical sciences, first had put forth them forward and after their study began to consider them the beginnings of all things ... Since, therefore, everything else was*

*explicitly compared to numbers throughout their essence, and numbers took first place in the whole of nature, they had recognized harmony and number as the basis of all things and all Universe."*

### 1.1.3. The contribution of Heraclitus to the development of the doctrine of *Harmony*

Starting from antiquity to the present day, Heraclitus remains one of the most popular philosophers in the history of philosophy. In 1961, on the recommendation of the World Peace Council, the 2500th anniversary of the birth of Heraclitus was celebrated. Such an anniversary is usually celebrated to commemmorate the history of some world-famous ancient cities or countries, but to do so for a person is rare and unusual.

Heraclitus believed that everything is constantly changing. The idea of *the eternal motion* was presented by Heraclitus in the bright image of the ever-flowing river (Fig. 1.2). The postulate on the universal variability of the world, one of the cornerstones of all dialectics, is compressed by Heraclitus in the famous formula: *"It is impossible to enter twice into the same river."*

Fig. 1.2. Heraclitus.

As Shestakov points out [10], *"in the aesthetics of Heraclitus, ontological understanding of harmony is at the forefront. Harmony is inherent, above all, the objective world of things, the cosmos itself what is inherent to the nature of art. It is characteristic that when Heraclitus wants to reveal the nature of harmony most clearly, he turns to fine arts. Best of all, Heraclitus illustrated the harmony of the Cosmos by the image of the lyre, in which the differently strained strings create a perfect harmony."*

But, in the aesthetics of Heraclitus, there is also a moment of evaluation. This is especially pronounced in the doctrine of two kinds of harmony: *"hidden"* and *"obvious"*. Heraclitus prefers the *"hidden"* Harmony. Widely known is the following saying of Heraclitus: *"The hidden harmony is stronger than the obvious."*

Cosmos, as the highest and most perfect beauty, is an example of the *Hidden Harmony*. Only at first glance the cosmos seems to be a chaos. In fact, a play of elements and events conceals *"the most beautiful harmony"*.

### 1.1.4. The musical harmony of Pythagoras and the music of the spheres

Pythagoreans made wonderful discoveries in music. Pythagoras found that the most pleasant to ear consonances are obtained only when the lengths of the strings, that produce these consonances, have ratios as the first natural numbers 1, 2, 3, 4, 5, 6, that is, 1:1, 1:2 (*unison* and *octave*), 2:3, 3:4 (*quint* and *quart*), 4:5, 5:6 (*thirds*), etc. The discovery he made (the *law of consonances*) shocked Pythagoras. It was this discovery that first pointed out the existence of numerical patterns in Nature, and it was this that served as a starting point in the development of *Pythagorean philosophy* and in the formation of their basic thesis: *"Everything is a Number."* Therefore, the day when Pythagoras discovered the *law of consonances*, was declared by the German physicist A. Sommerfeld the birthday of theoretical physics.

The discovery of mathematical regularities in musical consonances was the first "experimental" confirmation of the Pythagorean doctrine of *Number*. From this moment, the *music* and the related doctrine of *Harmony* began to occupy a central place in the *Pythagorean system of knowledge*. The idea of musical relations soon acquired the "cosmic scales" among the Pythagoreans and grew into the idea of *Universal Harmony*.

The Pythagoreans began to assert that the entire Universe is organized on the basis of simple numerical relationships and that the moving planets demonstrate *"the music of the heavenly spheres"* and ordinary music is merely a reflection of *Universal Harmony* prevailing everywhere. Thus, music and astronomy had been reduced by the Pythagoreans to the analysis of numerical relations, that is, to *arithmetics* and *geometry*. All four MATEMs (*arithmetics, geometry, harmonics* and *spherics*) began to be considered mathematical and called by one word — *"mathematics."*

### 1.1.5. Once again about the term of the *Mathematics of Harmony*

As mentioned in the preface, for the first time, the term *Mathematics of Harmony* was introduced in a short article "Harmony of spheres", published in The Oxford Dictionary of Philosophy [103]. This doctrine, often attributed to Pythagoras, leads to the unification of mathematics, music, and astronomy. Its essence is the fact that the celestial solids, being huge objects, must produce music during their movement. The perfection of the heavenly world requires that this music must be *harmonious*; it is hidden from our ears only because it is always present. The *Mathematics of Harmony* was a Central Discovery of Immense Importance for the Pythagoreans.

Thus, the concept of the *Mathematics of Harmony* in [103] is associated with the "Harmony of spheres", which was also called the *Harmony of the World* or *world music*. The "Harmony of spheres" is an ancient and medieval doctrine about the musical and mathematical structure of the Cosmos, which goes back to the Pythagoras and Plato philosophical tradition.

Another mention about the *Mathematics of Harmony*, applied to ancient Greek mathematics, is found in the book by Vladimir Dimitrov *A New Kind of Social Science. The Study of*

*Self-Organization of Human Dynamics*, published in 2005 [54]. Here is a quote from this book:

> *"The prerequisite for Harmony for the Greeks was expressed with the phrase "nothing superfluous." This phrase contained mysterious positive qualities, which became the object of the study of the best minds. Thinkers, such as Pythagoras, sought to unravel the mystery of Harmony as something unspeakable and illuminated by mathematics."*
>
> ***The Mathematics of Harmony***, *studied by the ancient Greeks, is still an inspiring model for modern scholars. Of decisive importance for this was the discovery of the quantitative expression Harmony, in all the amazing variety and complexity of Nature, through the golden section* $\Phi = (1 + \sqrt{5})/2$ *that is approximately equal to 1.618. The golden section is described by Euclid in his Elements: "It is said that a straight line can be divided in extreme and mean ratio, when, the entire line so refers to the most part, like most to lesser."*

It is important to emphasize one more aspect in the book [54] — the concept of *Mathematics of Harmony* is directly associated with the *golden section*, the most important mathematical discovery of the ancient science in the field of the *Mathematical Harmony*, which at that time was called the *"division of a segment in the extreme and mean ratio."*

## 1.2. The Golden Section in Euclid's *Elements*

Euclid's *Elements* is the greatest mathematical work of the ancient Greeks. Currently, every school student knows the name of Euclid, who wrote the most significant mathematical work of the Greek epoch, Euclid's *Elements*. This scientific work was created by him in the third century BC and contains the foundations of ancient mathematics: *elementary geometry, number theory, algebra, the theory of proportions and relations, methods for the calculating areas and volumes*, etc. In this work, Euclid summed the development of Greek mathematics and created a solid foundation for its development (Fig. 1.3).

The information about Euclid is extremely scarce. For the most reliable information about Euclid's life, it is customary to relate the little data that is given in Proclus' *Commentaries to the first book*

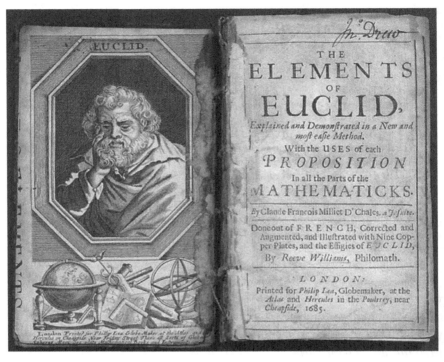

Fig. 1.3.  Euclid.

*of Euclid's Elements.* Proclus points out that Euclid lived during the time of Ptolemy I Soter, because Archimedes, who lived under Ptolemy I, mentions Euclid. In particular, Archimedes says that Ptolemy once asked Euclid if there is a shorter way of studying geometry than *Elements*; and Euclid replied that *there was no royal path to geometry.*

Plato's disciples were teachers of Euclid in Athens and in the reign of Ptolemy I (306–283 BC). Euclid taught at the newly founded school in Alexandria. Euclid's *Elements* had surpassed the works of his predecessors in the field of geometry and for more than two millennia remained the main work on elementary mathematics. This unique mathematical work contained most of the knowledge on geometry and arithmetic of Euclid's era.

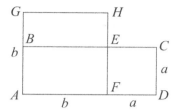

Fig. 1.4. Division of a segment in the extreme and mean ratio.

### 1.2.1. Proposition II.11 of Euclid's *Elements*

In Euclid's *Elements*, we find a task that later played an important role in the development of science. It was called *Dividing a segment in the extreme and mean ratio*. In the *Elements*, this task occurs in two forms. The first form is formulated in Proposition II.11 of Book II [32].

**Proposition II.11 (see Fig. 1.4).** *The given segment AD is divided into two unequal segments AF and FD so that the area of the square AGHF, constructed on a larger segment AF, must equal to the area of the rectangle ABCD, constructed on a larger (AF) and smaller (FD) segments.*

Let's try understanding the essence of this task by using Fig. 1.4.

If we denote the length of the larger segment $AF$ through $b$ (it is equal to the side of the square $AGHE$), and the side of the smaller segment through $a$ (it is equal to the vertical side of the rectangle $ABCD$), then the condition of the Proposition II.11 can be written in the form

$$b^2 = a \times (a + b). \tag{1.1}$$

### 1.2.2. The second form of the task of the division of segment in the extreme and mean ratio

The second form follows from the first form, given by (1.1), if we will make the following transformations. Dividing both sides of the expression (1.1) first by $a$, and then by $b$, we obtain the following

Fig. 1.5. Division of a segment in extreme and mean ratio (the golden section).

proportion:

$$\frac{b}{a} = \frac{a+b}{b}. \tag{1.2}$$

The proportion (1.2) has the following geometric interpretation (Fig. 1.5). We divide the segment $AB$ by the point $C$ for the two inequal segments $AC$ and $CB$ in such a manner that the bigger segment $CB$ so refers to the smaller segment $AC$, as how the whole segment $AB$ refers to the bigger segment $CB$, that is,

$$\frac{AB}{CB} = \frac{CB}{AC}. \tag{1.3}$$

This is the definition of the "golden section", which is used in modern science.

We denote the proportion (1.3) by $x$. Then, taking into consideration that $AB = AC + CB$, the proportion (1.3) can be written in the following form:

$$x = \frac{AC + CB}{CB} = 1 + \frac{AC}{CB} = 1 + \frac{1}{\frac{CB}{AC}} = 1 + \frac{1}{x},$$

from which the following algebraic equation for calculating the desired ratio $x$ follows:

$$x^2 - x - 1 = 0. \tag{1.4}$$

It follows from the "physical meaning" of the proportion (1.3) that the desired solution of the equation (1.4) must be a positive number, from which it follows that the solution of task of "*dividing a segment in extreme and mean ratio*" [32] is the positive root of equation (1.4), which we denote by $\Phi$:

$$\Phi = \frac{1 + \sqrt{5}}{2}. \tag{1.5}$$

Fig. 1.6.  Phidias (490–430 BC).

This is the famous irrational number that has many delightful names: *golden section, golden number, golden proportion, divine proportion.*

The algebraic equation (1.4) is often called the *equation of the golden proportion.*

Note that on the segment $AB$, there is one more point $D$ (Fig. 1.6), which divides $AB$ in the *golden section*, because

$$\frac{AB}{AD} = \frac{AD}{DB} = \frac{1 + \sqrt{5}}{2}.$$

### 1.2.3.  Comments of Mordukhai-Boltovsky, concerning the golden section

Euclid's *Elements* are translated into many languages of the world. The most authoritative edition of Euclid's work in Russian is Euclid's *Elements* in translation and with comments by the Russian geometer D.D. Mordukhai-Boltovsky [104–106]. It is interesting to read the following Mordukhai-Boltovsky comments about the *golden section*:

*"Now let's see what place takes the golden section in Euclid's Elements. First of all, it should be noted that it is realized in two forms, the difference between which is almost imperceptible for us, but it was very significant in the eyes of the Greek mathematicians of the 5th–6th centuries BC. The first form, the prototype of which we saw in Egypt, is*

*in Book II of the Elements, namely in Proposition II.11, together with the sentences 5 and 6 that introduce it; here the golden section is defined as such in which the area of a square, built on a bigger segment, equals to the area of a rectangle, built on the entire straight line and the smaller segment.*

*The second form we have in Definition 3 of Book VI, where the golden section is determined by the proportion, like the whole straight line to the larger segment, and the larger segment to the smaller one, and is called the division in extreme and mean ratio; in this form the golden section could only be known from Eudoxus' time ...*

*In Book XIII, the golden section appears in both of these forms, namely in the first form in Propositions 1–5 and in the second form in Propositions 8–10 ... Moreover, Proposition 2 of Book XIII is essentially equivalent to the geometric construction of Proposition 11 of Book II.*

*All this allows to think that Propositions 4, 7, 8 of Book II and Propositions 1–5 of Book XIII represent the remains of one of the most ancient documents in the history of Greek geometry, which most probably goes back to the first half of the 5th century and originated in the Pythagorean school on the basis of the material that was brought from Egypt...*"

We can draw the following conclusions from these comments:

(1) First of all, Euclid's *Elements* contains not just one (Proposition II.11), but at least two different formulations of the *golden section*. As follows from the comments of Mordukhai-Boltovsky, Euclid widely uses in his *Elements* both the first form (Proposition II.11 of Book II and Propositions 1–5 of Book XIII) and the second form as a representation of the *golden section* in the form of a proportion (Proposition 3 of Book VI and Propositions 8–10 of Book XIII).

(2) In the *golden section* (Proposition II.11), Mordukhai-Boltovsky sees the "Egyptian trace" and clearly hints at Pythagoras, who spent 22 years in Egypt and brought from there a huge amount of Egyptian mathematical knowledge, including the "Pythagoras theorem" and the *golden section*. Hence, it follows from Mordukhai-Boltovsky's comments that Mordukhai-Boltovsky did not doubt that not only Euclid, but also Pythagoras and Plato (who was a Pythagorean), and also the ancient Egyptians knew about the *golden section* and

widely used it (in what follows, we will demonstrate this by analyzing the geometric model of the Cheops Pyramid as an example).

### 1.2.4. The origin of the term of the golden section

The approximate value of the *golden proportion* is the following:

$$\Phi \approx 1.61803\,39887\,49894\,84820\,45868\,34365\,63811\,77203\,09180 \ldots.$$

Do not be surprised by this number! Do not forget that this number is irrational! In our book, we will use the following approximate value of $\Phi$: $\Phi \approx 1.618$ or even $\Phi \approx 1.62$.

It is this amazing number, possessing unique algebraic and geometric properties, that has become an aesthetic canon of ancient Greek and Renaissance arts.

Who introduced the term golden section? Sometimes, the introduction of this name ("section aurea") is attributed to Leonardo da Vinci. However, there is an opinion that the great Leonardo was not the first. According to the statement of Edward Soroko [4], this term goes back to Ptolemy's book *About Harmony*. However, in the book [32] on the history of the "golden number", it is stated that the German mathematician Martin Om first introduced the term "goldener Schnitt" in 1835 in the book *Die reine elementar mathematik.*

The designation of the *golden proportion* by the Greek letter $\Phi$ (the number $\Phi$) is not accidental. This letter is the first letter in the Greek name of the famous Greek sculptor Phidias (Greek: Φειδας), who widely used the *golden section* in his sculptural works. Recall that Phidias (480–430 BC) was one of the most significant and authoritative masters of ancient Greek sculpture of the classics era (Fig. 1.6).

As a sculptor, he became famous for having created two grandiose (gold and ivory) statues: Athens Parthenos for the Parthenon at the Acropolis (446–438 BC) and Olympus Zeus (for the Temple of Zeus at Olympia, circa 430 BC), which were considered as one of the "Seven Wonders of the World". For all the monumentality of these

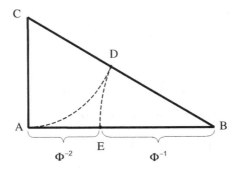

Fig. 1.7. The method of geometric construction of the golden section.

sculptures, unprecedented in size for Greece of that time, they were characterized by a strict balance and harmony of plastic contrasts, based on the golden section, which was the essence of the classical style in the period when it was flourishing the highest.

The essence of the method in Fig. 1.7 is as follows. We construct the right-angular triangle $ABC$ with the sides $AB = 1$, $AC = \frac{1}{2}$. Then, in accordance with the "Pythagoras theorem", we calculate the side $CB = \sqrt{1 + (\frac{1}{2})^2} = \frac{\sqrt{5}}{2}$. Drawing the arc $AD$ with the center at the point $C$ before the intersection with the segment $CB$ at the point $D$, we obtain the segment $BD = CB - CD = \frac{\sqrt{5}-1}{2} = \Phi^{-1}$. By drawing the arc $BD$ with the center at the point $B$ before its intersection with the segment $AB$ at the point $E$, we obtain the division of the segment $AB$ at the point $E$ with the golden section, because

$$\frac{AB}{EB} = \frac{EB}{AE} = \Phi \quad \text{or} \quad AB = 1 = EB + AE = \Phi^{-1} + \Phi^{-2}.$$

## 1.3. Proclus Hypothesis and New View on Classic Mathematics and Mathematics of Harmony

### 1.3.1. For what purpose did Euclid write his *Elements*?

At first glance, it seems that the answer to this question is very simple: Euclid's main goal was to set forth the main achievements of Greek mathematics during 300 years, preceding Euclid, by using the "axiomatic method". Indeed, Euclid's *Elements* is the main work in Greek science, devoted to the axiomatic description of geometry and mathematics. This view on *Elements* is the most common one in modern mathematics.

However, besides the "axiomatic" point of view, there is another point of view on the motives that guided Euclid in writing the *Elements*. This point of view was described by the Greek philosopher and mathematician Proclus Diadoch (412–485 AD), one of the best commentators of Euclid's *Elements*.

First of all, a few words about Proclus (Fig. 1.8). Proclus was born in Byzantium in the family of a wealthy lawyer from Lycia. Intending to follow in the footsteps of his father, he left for Alexandria in his teens, where he studied at first rhetoric, then became interested in philosophy and became a disciple of the Neoplatonist Olympiodorus the Younger. It was here that Proclus began to study the logical treatises of Aristotle. At the age of 20, Proclus returned to Athens, where Plutarch of Athens headed the Platonic Academy. At the age of 28, Proclus wrote one of his most important works, *Commentary on Plato's "Timaeus"*. About 450 AD, Proclus becomes the Head of the Platonic Academy.

Among Proclus' mathematical works, the most famous is the *Commentary on the first Book of Euclid's Elements*. In this commentary, he puts forward the following unusual hypothesis, which

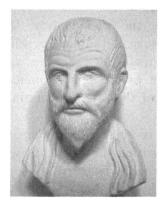

Fig. 1.8. Proclus.

is called the *Proclus hypothesis*. Its essence is as follows. As we know, Book XIII, that is, the final book of the *Elements*, is devoted to the presentation of the theory of the five regular polyhedra, which played a dominant role in *Plato's cosmology* and in modern science are known under the name of the *Platonic solids*. Proclus draws particular attention to this circumstance. As Edward Soroko points out in [4], according to Proclus, Euclid created the *Elements* allegedly not for the purpose of expounding geometry as such, but to give a complete systematized theory of constructing the *five Platonic solids*; in addition, he described here some of the latest achievements of Greek mathematics."

### 1.3.2. The significance of the Proclus hypothesis for the development of mathematics

The main conclusion from the Proclus hypothesis consists in the fact that Euclid's *Elements*, the greatest Greek mathematical work, was written by Euclid under the direct influence of the Greek "idea of Harmony", which was associated with the *Platonic solids*.

Thus, the Proclus hypothesis makes it possible to suggest that the well-known doctrines in ancient science *"Pythagorean doctrine of the numerical harmony of the Universe"* and *Plato's cosmology*, based on regular polyhedra, were embodied in Euclid's *Elements*, the greatest mathematical work of the ancient Greek mathematics. From this point of view, we can consider Euclid's *Elements* as the first attempt in creating the *Mathematical Theory of the Universal Harmony*, which was associated in ancient science with the *Platonic solids*. This was the main idea of the ancient Greek science! This is the main secret of Euclid's *Elements*, which leads to the revision of the history of the origin of mathematics, starting since Euclid.

Unfortunately, the original *Proclus hypothesis*, concerning the true Euclid's goals in writing the *Elements*, was ignored by modern historians of mathematics, which led to a distorted view of the structure of mathematics and the whole mathematical education. This is one of the main "strategic mistakes" in the development of mathematics.

The *Proclus hypothesis* had a great influence on the development of science and mathematics. In the 17th century, Johannes Kepler, by developing Euclid's ideas, built the *Cosmic Cup*, the original model of the *Solar system*, based on the *Platonic solids*.

In the 19th century, the eminent mathematician Felix Klein (1849–1925) proposed that the *icosahedron*, one of the most beautiful *Platonic solids*, is the main geometric figure of mathematics, which makes it possible to unite all the most important branches of mathematics: *geometry*, *Galois theory*, *group theory*, *invariant theory*, and *differential equations* [113]. Klein's idea did not get further support in the development of mathematics, which can also be considered as another "strategic mistake".

### 1.3.3. The Proclus hypothesis and "key" problems of ancient mathematics

As it is well known, the Russian mathematician academician Kolmogorov in his book [102] singled out two "key" problems, which stimulated the development of mathematics at the stage of its origin, the problems of *counting* and *measurement*. However, another "key" problem, which emerged from the *Proclus hypothesis* is the *harmony problem* which was connected with *Platonic solids* and the *golden section*, one of the most important mathematical discoveries of ancient mathematics (Proposition II.11 of Euclid's *Elements*). It was this problem that Euclid laid in the foundation of *Elements*. According to Proclus, the main purpose of *Elements* was to create the geometric theory of the *Platonic solids*, which in *Plato's cosmology* was expressed as *Universal Harmony*. This idea leads to a new view on the history of mathematics, as presented in Fig. 1.9.

The approach, shown in Fig. 1.9, is based on the following reasoning. Already, at the stage of origin of mathematics, a number of important mathematical discoveries were obtained; they fundamentally influenced the development of mathematics and all science on the whole. The most important of them are as follows:

(1) **The positional principle of numbers representation.** This was discovered by Babylonian mathematicians in the second

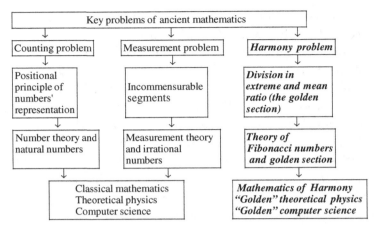

Fig. 1.9. "Key" problems of ancient mathematics and new directions in mathematics, theoretical physics and computer science.

millennium BC and embodied by them in the *Babylonian positional numeral system with the base of 60*. This important mathematical discovery underlies all subsequent positional numeral systems, in particular, the *decimal system*, the basis of mathematical education, and the *binary system*, the basis of modern computers. This discovery, ultimately, led to the formation of the concept of *natural numbers*, one of the most important concepts, underlying mathematics.

(2) **Incommensurable segments**. This discovery, made by Pythagoreans, led to a rethinking of the early Pythagorean mathematics, which was based on the *principle of magnitudes' commensurability*, and to the introduction of *irrational numbers*, the second (after *natural numbers*) fundamental concept of mathematics. Ultimately, these two concepts (the natural and irrational numbers) were used as the basis for *Classical Mathematics*.

(3) **Division of a segment in the extreme and mean ratio (the golden section)**. For the first time, the description of this discovery was given in Euclid's *Elements* (Proposition II.11). This proposal was introduced by Euclid with the goal of creating a complete geometric theory of *Platonic solids* (in particular, the

*dodecahedron*), whose presentation was given in the final (Book XIII) book of Euclid's *Elements*.

The above approach (Fig. 1.9) leads to the conclusion, which may be unexpected for many mathematicians. It turns out that in parallel with the *Classical Mathematics*, starting with the ancient Greeks, another mathematical direction developed, the *Mathematics of Harmony*, which, like to the *Classical Mathematics*, goes back to Euclid's *Elements*, but focuses its attention not only on the *axiomatic approach*, but also on the geometric *"problem of dividing a segment in the extreme and mean ratio"* (Proposition II.11) and on the theory of regular polyhedra, set forth in Book XIII of Euclid's *Elements*.

At the same time, many outstanding thinkers and scientists took part in the development of the *Mathematics of Harmony* for several millennia: Pythagoras, Plato, Euclid, Fibonacci, Pacioli, Kepler, Cassini, Binet, Lucas, Klein, and, in the 20th century, Vorobyov and Hoggatt.

### 1.3.4. How did Euclid use the golden section in his *Elements*?

The question arises: Why did Euclid introduce the various forms of "the problem of dividing of a segment in the extreme and mean ratio" (the *golden section*) that are found in Books II, VI and XIII of the *Elements*? To answer this question, we return to the *Platonic solids* (see Fig. 1.4).

As it is well known, the faces of *Platonic solids* can be only three kinds of regular polygons: *regular triangle* (*tetrahedron, octahedron, and icosahedron*), *square* (*cube*), and a *regular pentagon* (*dodecahedron*). In order to design the *Platonic solids*, we must, first of all, be able to geometrically (that is, by using a ruler and a compass) construct the faces of the *Platonic solids*. Euclid did not have problems with the construction of a regular or equilateral triangle and a square, but he encountered some difficulties in constructing the regular pentagon, which underlies the *dodecahedron* (Fig. 1.10).

Fig. 1.10.  Dodecahedron.

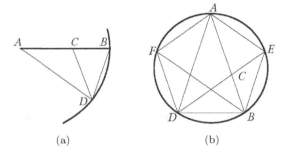

(a)                              (b)

Fig. 1.11.  The "golden" (a) isosceles triangle and (b) the pentagon.

It is for this purpose that Euclid in Book II had introduced the *golden section* (Proposition II.11), represented in the *Elements* in two forms. By using the *golden section*, Euclid first constructs a "golden" isosceles triangle, whose angles at the base are equal to twice of the angle at the vertex (Fig. 1.11(a)).

To do this, the segment $AB$ is separated by the point $C$ in the *golden section*. Then the circle with the center at the point $A$ and the radius $AB$ is drawn. After that, the span of the compass is chosen equal to the segment $AC$ and on the circle (with the aid of a compass) is marked the point $D$, such that $AC = CD$. Then, by using the ruler, the segments $AD$, $CD$ and $BD$ are drawn. The triangle $ABD$ has the property that the angles $B$ and $D$ at its base $BD$ are equal to twice the angle at its vertex $A$, and the segments $CD$ and $BD$ prove to be equal.

Now, let's pass to the construction of the *regular pentagon* (Fig. 1.10). To do this, we start with the triangle $ABD$, constructed in Fig. 1.11(a). Draw the circle through the points $A$, $B$ and $D$ (Fig. 1.11(b)). After that, we divide the angle $ABD$ in half and draw a segment $DE$ to its intersection with the circle at the point $E$. Note that this segment intersects at the point $C$ with the segment $AB$, dividing it in the *golden section*. Similarly, we find the point $F$ on the circle and then find the regular pentagon $AEBDF$. Then we have only one step to the geometric construction of the *dodecahedron* (Fig. 1.10), which is one of the most important regular polyhedra, the symbol of the Universal Harmony in Plato's cosmology.

## 1.4. Some Simplest Mathematical Properties of the Golden Ratio

### 1.4.1. The simplest algebraic identities for the golden ratio

What is this "miracle" of nature and mathematics, in which one's interest does not fade with time, and, on the contrary, grows with each century? To answer this question, we suggest the readers to bull together all their mathematical knowledge and immerse themselves in the world of mathematics. Only in this way we can enjoy the wonderful mathematical properties of this unique phenomenon and, through these mathematical properties and identities, appreciate all the beauty and harmony of the *golden ratio* $\Phi = \frac{1+\sqrt{5}}{2} \approx 1.618$.

Let us begin by establishing the simplest algebraic identities of the golden ratio. To this end, we represent equation (1.4) in the following form:

$$x^2 = x + 1. \tag{1.6}$$

If the root $\Phi$ (the golden proportion) is substituted instead of $x$ in equation (1.6), then we get the following remarkable identity for the golden proportion:

$$\Phi^2 = \Phi + 1. \tag{1.7}$$

If we divide the left and right parts of the identity (1.7) on $\Phi$, we get the following expression for $\Phi$:

$$\Phi = 1 + \frac{1}{\Phi}, \qquad (1.8)$$

which can be represented as follows:

$$\Phi - 1 = \frac{1}{\Phi}. \qquad (1.9)$$

Let us analyze the identity (1.9). It is known that every number $a$ has the inverse number $1/a$. For example, the fraction of 0.1 is the number inverse to 10. The traditional algorithm for obtaining the inverse number from the original number $a$ consists in dividing the number 1 by the number $a$. This is a rather complicated procedure. Try, for example, by dividing to get the number, inverse to the number $a = 357821$. This can only be done with a modern computer.

Consider now the golden proportion $\Phi = \frac{1+\sqrt{5}}{2}$. How to get the inverse to it number $\frac{1}{\Phi}$? The identity (1.9) gives a very simple answer to this question. To do this, it is enough to subtract the unit 1 from the golden proportion $\Phi = \frac{1+\sqrt{5}}{2}$.

But we get even more "aesthetic pleasure" if we carry out the following transformations over the identity (1.9). First, let's multiply both sides of the identity (1.9) by the golden ratio $\Phi$, and then let's divide them by $\Phi$. As a result, we get two new identities:

$$\Phi^3 = \Phi^2 + \Phi \qquad (1.10)$$

and

$$\Phi = 1 + \Phi^{-1}. \qquad (1.11)$$

If we now continue multiplying the terms of the identity (1.10) by $\Phi$ and dividing the terms of the identity (1.11) by $\Phi$ and will continue this process *ad infinitum*, then we get the following elegant identity, connecting the degrees of the *golden proportion*:

$$\Phi^n = \Phi^{n-1} + \Phi^{n-2}, \quad n = 0, \pm 1, \pm 2, \pm 3, \ldots \qquad (1.12)$$

The identity (1.12) can be verbally expressed as follows: "*Any whole degree of the golden proportion is equal to the sum of the two previous ones.*"

### 1.4.2. "Golden" geometric progression

Consider a sequence of degrees of the *golden proportion*, that is,

$$\{\dots, \Phi^{-n}, \Phi^{-(n-1)}, \dots, \Phi^{-2}, \Phi^{-1}, \Phi^0 = 1, \Phi^1, \Phi^2, \dots, \Phi^{n-1}, \Phi^n, \dots\}. \tag{1.13}$$

The sequence (1.13) has very interesting mathematical properties. On the one hand, the sequence (1.13) is a "geometric progression", in which each number is equal to the previous number, multiplied by the number $\Phi$, called the denominator of the geometric progression, that is,

$$\Phi^n = \Phi \times \Phi^{n-1}. \tag{1.14}$$

On the other hand, in accordance with (1.12), the sequence (1.13) has an "*additive*" *property* because each member of the progression (1.13) is the sum of the two preceding ones. We note that the property (1.12) is characteristic only for the geometric progression with the denominator $\Phi$ and such geometric progression is called the *golden progression.*

Since every geometric progression of the type (1.13) corresponds to a certain logarithmic spiral in geometry, many authors believe that the property (1.12) distinguishes the *golden progression* (1.13) among other geometric progressions and is the reason for the wide distribution of the "*golden*" *logarithmic spiral* in the forms and structures of living nature.

### 1.4.3. Representation of the golden proportion in the "radicals"

Let us now consider once again the identity (1.7). If we take the square root from the right and left sides of the identity (1.7), then we get the following expression for $\Phi$:

$$\Phi = \sqrt{1 + \Phi}. \tag{1.15}$$

If now in the right side of the expression (1.15), we substitute instead $\Phi$ the expression, given by (1.15), we get the following expression:

$$\Phi = \sqrt{1 + \sqrt{1 + \Phi}}. \tag{1.16}$$

If we do this substitution infinite number of times, then we will get one more remarkable representation of the *golden proportion* in the "radicals":

$$\Phi = \sqrt{1 + \sqrt{1 + \sqrt{1 + \sqrt{1 + \cdots}}}}. \tag{1.17}$$

## 1.5. The Golden Ratio and Chain Fractions

### 1.5.1. General information about chain fractions

Chain fraction (or continued fraction) is a mathematical expression of the form

$$x = [a_0; a_1, a_2, a_3, \ldots] = a_0 + \cfrac{1}{a_1 + \cfrac{1}{a_2 + \cfrac{1}{a_3 + \cdots}}}, \tag{1.18}$$

where $a_0$ is an integer and all others $a_n(n = 1, 2, 3, \ldots)$ are positive integers, that is, natural numbers, including the number 0. Any real number can be represented as a continued fraction (finite or infinite). A number is represented by a finite continued fraction if and only if it is rational. A number is represented by a periodic continued fraction if and only if it is a quadratic irrationality.

The theory of continued fractions goes back to ancient mathematics. It is believed that ancient mathematicians had been able to represent the relations of incommensurable quantities in the form of a chain of successive suitable relations by obtaining this chain through the *Euclidean algorithm*.

## 1.5.2. Representation of the "golden proportion" in the form of a chain fraction

We now represent the golden ratio $\Phi$ in the form of a chain fraction (1.18). To this end, we use the identity (1.8). If we substitute the value, given by the same expression (1.8), instead of $\Phi$ in the right-hand side of (1.8), then we get the representation $\Phi$ in the form of the following "multi-storey" fraction:

$$\Phi = 1 + \frac{1}{1 + \frac{1}{\Phi}}. \tag{1.19}$$

If we continue this substitution into the right-hand side of (1.19) *ad infinitum*, then as a result, we get a "multi-storey" fraction with an infinite number of "floors":

$$\Phi = 1 + \frac{1}{1 + \frac{1}{1 + \frac{1}{1 + \frac{1}{1 + \cdots}}}}. \tag{1.20}$$

It turns out that the expression (1.20) has a profound mathematical sense. The Russian mathematicians Aleksandr Khinchin [107] and Nikolay Vorobyov [8] had drawn attention to the fact that the expression (1.20) distinguishes the golden proportion among other irrational numbers because, according to (1.20), the golden ratio is most slowly approximated by rational fractions. This means that, in terms of chain fractions, the golden proportion is a *unique irrational number*.

We now find suitable fractions for the golden ratio. To this end, by using (1.20), we approximate the golden proportion (1.20) with the following appropriate fractions:

$$1 = \frac{1}{1} \quad \text{(first approximation)},$$

$$1 + \frac{1}{1} = \frac{2}{1} \quad \text{(second approximation)},$$

$$1 + \frac{1}{1 + \frac{1}{1}} = \frac{3}{2} \quad \text{(third approximation)},$$

$$1 + \frac{1}{1 + \frac{1}{1 + \frac{1}{1}}} = \frac{5}{3} \quad \text{(fours approximation)}.$$

Continuing this process, we find a sequence of suitable fractions for the *golden proportion*, which is a sequence of ratios of the neighboring Fibonacci numbers:

$$\frac{1}{1}, \frac{2}{1}, \frac{3}{2}, \frac{5}{3}, \frac{8}{5}, \frac{13}{8}, \frac{21}{13}, \cdots \rightarrow \Phi = \lim_{n \to \infty} \frac{F_n}{F_{n-1}} = \frac{1 + \sqrt{5}}{2}. \qquad (1.21)$$

As it is known, the number sequence (1.21) expresses nothing but the famous law of phyllotaxis, according to which Nature constructs pinecones, pineapples, cacti, sunflower heads, and other botanical objects. In other words, Nature uses the unique mathematical property of the golden ratio, given by (1.20) and (1.21), in its remarkable creations!

Finally, a few words about the "aesthetic pleasure" of the identities (1.17) and (1.20) obtained above. Every mathematician intuitively seeks to express his mathematical results in the simplest and compact form. If such a form could be found, then this result presents "aesthetic pleasure" to the mathematician. In this respect (the desire for an "aesthetic" expression of mathematical results), mathematical creativity is similar to the work of a composer or poet, whose main task is to obtain perfect musical or poetic forms that provide "aesthetic pleasure." Note that formulas (1.17) and (1.20) give us "aesthetic pleasure" and cause an unconscious sense of rhythm and harmony when we begin to think about the infinite repeatability of the same simple mathematical elements in the formulas for $\Phi$, given by (1.17) and (1.20).

## 1.6. Equations of the Golden Proportion of the *N*th Degree

Usually, the main question that is solved in the theory of algebraic equations is to find the roots of a given algebraic equation. Now, we will put another question. We know the simplest algebraic equation (1.4), whose root is the golden proportion. We set forth the following question: Are there algebraic equations of higher degrees, whose roots are the golden proportion? If so, what form do they have?

To answer this question, we will carry out the following reasoning, by starting from the equation the golden proportion (1.4).

We multiply both sides of the equation of the golden proportion $x^2 = x + 1$ by $x$; as a result, we obtain the following expression:

$$x^3 = x^2 + x. \tag{1.22}$$

The equation $x^2 = x + 1$ can be written as follows:

$$x = x^2 - 1. \tag{1.23}$$

Substituting the expression (1.23) instead of $x$ into the expression (1.22), we obtain the following algebraic equation of the third degree:

$$x^3 = 2x^2 - 1. \tag{1.24}$$

On the other hand, if we substitute into the expression (1.22) instead $x^2$ the expression $x^2 = x + 1$, we obtain another algebraic equation of the third degree:

$$x^3 = 2x + 1. \tag{1.25}$$

Thus, we obtained two new algebraic equations of the third degree. Let's prove, for example, that the root of the equation (1.24) is the *golden proportion* $\Phi = \frac{1+\sqrt{5}}{2}$. To prove this, we substitute the *golden proportion* in the left and right sides of the equation (1.24) and then verify that the left and right sides of this equation coincide. Indeed, by using the identity (1.12), we obtain the following result for the left-hand side:

$$\Phi^3 = \Phi^2 + \Phi = \frac{3 + \sqrt{5}}{2} + \frac{1 + \sqrt{5}}{2} = \frac{4 + 2\sqrt{5}}{2}.$$

On the other hand, for the right-hand side of the equation (1.24), we have

$$2\Phi^2 - 1 = \frac{2(3 + \sqrt{5})}{2} - 1 = \frac{4 + 2\sqrt{5}}{2}.$$

Thus, the above argument proves that the *golden proportion* $\Phi$ is indeed the root of equation (1.24), that is, the algebraic equation (1.24) is the *"golden" equation*.

In exactly the same way, one can prove that equation (1.25) is also the *"golden" equation*.

We now derive the expression for the "golden" algebraic equation of the fourth degree. For this, we multiply both sides of (1.22) by $x$; as a result, we obtain the following equality:

$$x^4 = x^3 + x^2. \tag{1.26}$$

Then we can use the expression $x^2 = x + 1$ and the expressions (1.24) or (1.25) for $x^3$. By substituting them into the expression (1.26), we obtain two new algebraic equations of the fourth degree, whose roots are the golden proportion:

$$x^4 = 3x^2 - 1, \tag{1.27}$$

$$x^4 = 3x + 2. \tag{1.28}$$

The analysis of equation (1.28) leads us to an unexpected result. It turns out that this equation describes the energy state of the butadiene molecule, the valuable chemical substance that is used in the production of rubber. The well-known American physicist, Nobel Prize winner Richard Feynman expressed his admiration for the equation (1.28) in the following words: *"What miracles exist in mathematics! According to my theory, the golden proportion of the ancient Greeks gives the minimum energy state of the butadiene molecule."*

This observation of the famous physicist immediately raises our interest in the equations of the golden proportion of higher degrees, which, perhaps, describe the energy states of molecules of other chemical substances. These equations can be obtained if we consider successively the equations of the following types: $x^n = x^{n-1} + x^{n-2}$. As an example, we can derive the following "golden" equations of higher degrees:

$$x^5 = 5x + 3 = 5x^2 - 2; \quad x^6 = 8x + 5 = 8x^2 - 3;$$

$$x^7 = 13x + 8 = 13x^2 - 5.$$

The analysis of these equations shows that the numerical coefficients on the right-hand side of these equations are nothing as *Fibonacci numbers*. It is easy to show that in the general case, the algebraic equations of the golden proportion of the $n$th degree are expressed as follows:

$$x^n = F_n x^2 - F_{n-2} = F_n x + F_{n-1}, \qquad (1.29)$$

where $F_n$, $F_{n-1}$, $F_{n-2}$ are classic Fibonacci numbers.

Thus, our simple reasoning led us to a small mathematical discovery: we found an infinite number of new *"golden" algebraic equations*, given by the general expression (1.29).

If we substitute the golden proportion $\Phi$ instead $x$ in equation (1.29), we get two identities that connect the golden proportion to the Fibonacci numbers:

$$\Phi^n = F_n \Phi^2 - F_{n-2} \qquad (1.30)$$

$$\Phi^n = F_n \Phi + F_{n-1}. \qquad (1.31)$$

It is easy to calculate that, for example, the 18th, 19th and 20th Fibonacci numbers are equal, respectively,

$$F_{18} = 2584, \quad F_{19} = 4181, \quad F_{20} = 6765. \qquad (1.32)$$

But then we can write the following "golden" algebraic equations of the 20th degree:

$$x^{20} = 6765 x^2 - 2584, \qquad (1.33)$$

$$x^{20} = 6765 x + 4181. \qquad (1.34)$$

It is hard to imagine that the roots of these algebraic equations of the 20th degree are equal to the golden proportion! But this follows from the above theory of the "golden" algebraic equations, given by the expression (1.29). And looking at equations (1.29), (1.33), (1.34), we can once again be surprised at the greatness of mathematics, which allows us to express complex scientific information in such a compact form.

## 1.7. Geometric Figures Associated with the Golden Section

### 1.7.1. The Golden rectangle

We begin our journey along the geometric properties of the golden section with a *golden rectangle*, which has the following geometric definition (Fig. 1.12). The "golden" rectangle is a geometric figure, in which the ratio of the larger side to the smaller one is equal to the golden ratio, that is,

$$\frac{AB}{BC} = \Phi = \frac{1 + \sqrt{5}}{2}.$$

Consider the case when $AB = \Phi$ and $BC = 1$. We find on the segments $AB$ and $DC$ the points $E$ and $F$, which divide the corresponding sides in the golden section. It is clear that if $AE = DF = 1$, then $AB - AE = \Phi - 1 = \frac{1}{\Phi}$.

We now connect the points $E$ and $F$ by the segment $EF$ and call it the *golden line*. The *golden line* divides the golden rectangle $ABCD$ into two rectangles, $AEFD$ and $EBCF$. Since all sides of the rectangle $AEFD$ are equal to each other, this rectangle is a square.

Let's consider now the rectangle $EBCF$. Since its larger side $BC = 1$ and the smaller $EB = \frac{1}{\Phi}$, it follows from here that their ratio $\frac{BC}{EB} = \Phi$ and, therefore, the rectangle $EBCF$ is the "*golden*" *rectangle* also. Thus, the *golden line* $EF$ divides the initial "golden"

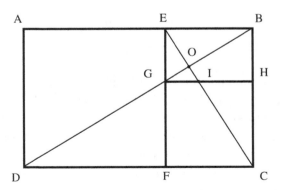

Fig. 1.12. The golden rectangle.

rectangle *ABCD* on the square *AEFD* and the new "golden" rectangle *EBCF*.

Let's conduct now the diagonals *DB* and *EC* of the "golden" rectangles *ABCD* and *EBCF*. It follows from similarity of the triangles *ABD, FEC* and *BCE* that the point *G* divides the diagonal *DB* in the *golden ratio*. Let's conduct now the new *golden line GH* in the *golden rectangle EBCF*. It is clear that the *golden line GH* divides the *golden rectangle EBCF* into the square *GHCF* and the new *golden rectangle EBHG*. By repeating this procedure many times, we get infinite sequence of squares and golden rectangles, which in the limit converge to the point O.

Note that such an infinite repetition of the same geometric figures, that is, squares and golden rectangles, causes in us an unconscious aesthetic sense of rhythm and harmony. It is believed that this circumstance is the reason why many rectangular objects with which we deal (matchboxes, lighters, books, suitcases) often have the form of *golden rectangle*.

### 1.7.2. A "double" square

As the Ukrainian researcher Nikolay Vasyutinsky points out in his book [24], *"many mathematical regularities, as they say, "had lain on the surface", they had been needed to be seen by a person with an analytical mind, logically thinking. And this was in abundance among the philosophers of the ancient world; because all of their scientific knowledge was built on the analysis of objects and phenomena and establishing a connection between them. In our time it is difficult to even imagine that it is possible to develop science without the use of experiment, but in fact such was the science of the Ancient World."*

It is possible that ancient mathematicians could find the *golden ratio*, by exploring the simplest rectangle with the ratio of sides 2:1, also called a *"double" square*, because it consists of two squares (Fig. 1.13).

Next, we present the arguments taken from Vasyutinsky's book [24]. If we calculate the diagonal *AC* of the *"double" square*

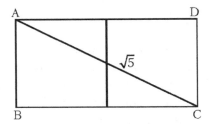

Fig. 1.13. A "double" square.

(Fig. 1.13), then, according to the Pythagoras theorem, it will be equal to $AC = \sqrt{5}$. If we now take the ratio of the sum of the segments $AC + CD$ to the larger side $AD$ of the *"double"* *square*, then we get the *golden ratio*, because $\frac{AC+CD}{AD} = \frac{1+\sqrt{5}}{2}$.

The *"double"* *square* in Fig. 1.13 consists of two right-angled triangles $ABC$ and $ADC$ with the aspect's ratio: $1 : 2 : \sqrt{5}$. It is reasonable to assume that such a triangle was well known to Pythagoras and could serve as the basis for the development of various mathematical ideas. First of all, we note that the sides of this triangle are connected by the Pythagoras theorem: $c^2 = a^2 + b^2$ or $(\sqrt{5})^2 = 5 = 1^2 + 2^2 = 1 + 4$.

The magnitude of the hypotenuse of such a triangle, equal to $\sqrt{5}$, could give rise to the discovery of *incommensurable segments* or *irrational numbers* (although, according to another legend, the Pythagoreans came to the discovery of incommensurable segments when they investigate the ratio of the diagonal of the square to its side). The ratio of the sides of this triangle is very simple: $a : b = 1 : 2$, $c : a = \sqrt{5} : 1$, $c : b = \sqrt{5} : 2$. However, one more relation follows from these ratios: $(a + c) : b = (1 + \sqrt{5}) : 2$ (the golden ratio).

Thus, a simple rectangular triangle with the ratio of sides: $1 : 2 : \sqrt{5}$, well known in the ancient world, could be the basis for the three great discoveries of Pythagora's *"theorem of squares"*, *golden ratio*, and *incommensurable segments*.

Of course, this is only the assumption of Nikolay Vasyutinsky [24], which wasn't devoid of inner logic. Possibly, Pythagoras' reasoning, which led him to the great mathematical discoveries, had other assumptions. Most likely, he came to the *"theorem of squares"*, based on the consideration of the rectangular triangle with the ratio of sides 3 : 4 : 5, which was well known from ancient times and was called "perfect", "sacred" or "Egyptian". Such a triangle looks like this: $3^2 + 4^2 = 5^2$. But we can only put forward the various hypotheses because Pythagoras did not leave behind any publications in written form.

### 1.7.3. The "golden" rectangular triangle (Kepler's triangle)

There is another rectangular triangle, associated with the *golden section* (Fig. 1.14). In such triangle, its sides, equal to $x$, $y$, $z$, are forming the geometric progression: $z : y = y : x$. The discovery of this triangle is attributed to Johannes Kepler.

In Kepler's triangle, the ratio of the legs $y : x$ is equal to the square root of the golden ratio, that is,

$$y : x = \sqrt{\Phi}. \tag{1.35}$$

In accordance with the "theorem of squares", the hypotenuse $z$ can be calculated from the formula:

$$z = \sqrt{x^2 + y^2}. \tag{1.36}$$

If we take $x = 1$, $y = \sqrt{\Phi}$, then

$$z = \sqrt{1 + \Phi} = \sqrt{\Phi^2} = \Phi. \tag{1.37}$$

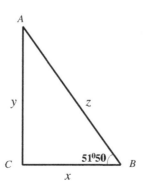

Fig. 1.14. Kepler's triangle.

The rectangular triangle, in which the sides relate as follows: $z : y : x = \sqrt{\Phi}$, called the *"golden" rectangular triangle* or *Kepler's triangle.*

There is quite a convincing hypothesis [24] the *"golden" rectangular triangle* (Fig. 1.14) was the main geometrical idea of Cheops pyramid (see below).

### 1.7.4. The "golden" ellipse

The golden ellipse is formed by using two rhombuses, *ACBD* and *ICJD*, inscribed in the ellipse (Fig. 1.15). The "golden" rhombuses, *ACBD* and *ICJD*, consist of the four "golden" rectangular triangles of the type *OCB* or *OCJ* (Fig. 1.15).

Consider the basic geometric relationships of the "golden" ellipse in Fig. 1.15. Let $AB = 2$ be the focal length of the ellipse. The definition of an ellipse implies the following relationship: $AC + CB = AG + GB$. On the other hand, there are the following relations linking the sides of the *"golden" right triangles **OCB** и **OCJ***:

$$OB : BC = 1 : \Phi, \quad OB : OC = 1 : \sqrt{\Phi};$$
$$OC : CJ = 1 : \Phi, \quad OC : OJ = 1 : \sqrt{\Phi},$$

where $\Phi$ is the golden proportion.

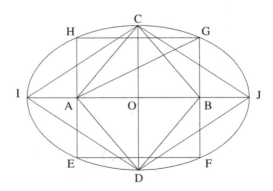

Fig. 1.15. The "golden" ellipse.

The following proportion follows from the similarity of the triangles $OCB$ and $OCJ$:

$$OC : CJ = OB : OC = OC : OJ = 1 : \sqrt{\Phi}. \qquad (1.38)$$

According to the Polish scientist, journalist and Egyptologist Jan Grzedzielski, the author of the remarkable book *Energy-Geometric Code of Nature* [20], the "golden" ellipse can be used as a geometric model for the dissemination of light in optical crystals, and then the proportions (1.38) express "harmonic" equilibrium in optical crystals, which create optimal conditions for photons to reach focuses with minimal energy losses.

**Pentagon and pentagram** are the two widely spread remarkable harmonic figures: *regular pentagon* and *pentagram* (Fig. 1.16). The word *pentagon* comes from the Greek word *pentagonon*.

If we conduct all the diagonals in the regular pentagon (Fig. 1.16), then we get a well-known *pentagonal star*, also called *pentagram* or *pentacle*. The name of the *pentagram* is derived from the Greek word *pentagrammon* (pente is five and grammon is the line).

It is proved that the points of intersection of diagonals in the regular pentagon are always the points of the golden section. In doing so, a new regular pentagon FGHKL is formed. In the new pentagon, we can draw new diagonals, the intersection of which

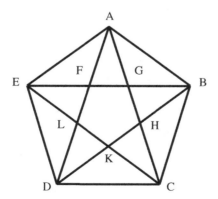

Fig. 1.16. Regular pentagon and pentagram.

forms the next regular pentagon and this process can be continued indefinitely. Thus, the regular pentagon ABCDE consists of an infinite number of regular pentagons, which are every time formed by the intersection points of the diagonals. This infinite repeatability of the same geometric figure creates a sense of rhythm and harmony, which is unconsciously fixed by our mind.

In the pentagon in Fig. 1.16, we can find a large number of the "golden" relations. For example, the ratio of the diagonal of the *pentagon* to its side is equal to the golden ratio.

We now consider the sequence of segments *FG, EF, EG, EB*. It is easy to show that they are connected by the *golden proportion*:

$$\frac{EF}{FG} = \frac{EG}{EF} = \frac{EB}{EG} = \Phi.$$

Note that the pentagram evoked special admiration among the Pythagoreans and was considered as their main identification sign.

### 1.7.5.  The golden bowl and the golden isosceles triangle

The regular pentagon and the pentagram in Fig. 1.16 include a number of remarkable geometric figures that have been widely used in works of fine art. In ancient art, there was a widely known so-called *golden bowl law* (Fig. 1.17); this law was used by antique sculptors and goldsmiths. The hatched part of the regular pentagon in Fig. 1.17 gives a schematic presentation of the *golden bowl*.

The pentagram (Fig. 1.16) consists of the five "golden" isosceles triangles (Fig. 1.18), each of which resembles the letter "A" (the "five intersecting A"). Each "golden" triangle in Fig. 1.18 has an acute angle $A = 36°$ at the vertex A and the two acute angles $D = C = 72°$ at the basis $DC$. The main feature of the "golden" isosceles triangle consists in the fact that the ratio of each hip $AC = AD$ to the basis $DC$ is equal to the golden proportion. By examining the "golden" isosceles triangle as part of the pentagram, the Pythagoreans were delighted when they found that the bisectrix $DH$ coincides with the

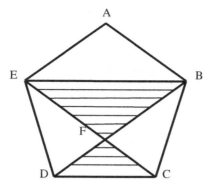

Fig. 1.17. The golden bowl.

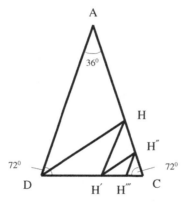

Fig. 1.18. The golden isosceles triangle.

diagonal of the regular pentagon $DB$ (Fig. 1.16) and divides the side $AC$ at the point $H$ in the golden section (Fig. 1.18). This creates a new golden triangle $DHC$. If we now draw the bisector of the angle $H$ to the point $H'$ and continue this process to infinity, then we obtain an infinite sequence of the "golden" isosceles triangles. As in the case with the "golden" rectangle (Fig. 1.13) and the regular pentagon (Fig. 1.16), the infinite appearance of the same geometric figure (the "golden" rectangular triangle) after the next bisector causes an aesthetic sense of rhythm and harmony.

## 1.7.6.  Decagon: the connection of the "golden section" with the number $\pi$

In mathematics, the most widely known are two mathematical constants: the number $\pi$, which expresses the ratio of the length of the circle to its diameter, and the *Naperian number e*, which is the basis of the so-called *natural logarithms*. The significance of these two most important mathematical constants in mathematical analysis is that they underlie the most important types of *elementary functions*, the well-known *trigonometric functions* (the number $\pi$), as well as the *exponential function* $e^x$, the *logarithmic function* $\log_e x$, finally, *hyperbolic functions* (the number $e$).

In mathematics, the following formula, derived by the famous mathematician Leonard Euler (1707–1783), is widely known:

$$e^{ix} = \cos x + i \sin x, \qquad (1.39)$$

where $e$ is the basis of natural logarithms, and $i = \sqrt{-1}$ (the *imaginary unit*). When $x = \pi$, we get the famous *Euler formula*:

$$e^{i\pi} + 1 = 0, \qquad (1.40)$$

which connects the numbers $\pi$ and $e$, the two main mathematical constants, "*dominating over the analysis*".

Not everyone can understand the *Euler formula* (1.40)! And looking at the identity (1.40), it is difficult to get rid from thought about the mystical character of the formula (1.40) based on the fundamental mathematical constants $\pi$ and $e$.

The golden proportion $\Phi$ also belongs to the category of fundamental mathematical constants. But then the question arises: is there any connection between these mathematical constants, for example, between the numbers $\pi$ and $\Phi$? The answer to this question is given by the analysis of a regular polygon called *decagon* (Fig. 1.19).

Let's consider a circle with the radius $R$ together with the *decagon* inscribed in the circle (Fig. 1.19). From the geometry, it is known that the side of the *decagon* is connected to the radius $R$ by the following

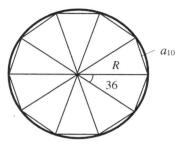

Fig. 1.19. The regular decagon.

formula:

$$a_{10} = 2R \sin 18°. \tag{1.41}$$

If we perform some trigonometric transformations based on the formulas well known to us from the course of school trigonometry, we get the following results:

(1) The side of the regular decagon, inscribed in the circle of the radius $R$, is equal to the greater part of the radius, divided by the golden ratio, that is,

$$a_{10} = R/\Phi. \tag{1.42}$$

(2) The *golden proportion* is connected to the number $\pi$ with the following relation:

$$\Phi = \frac{1 + \sqrt{5}}{2} = 2 \cos 36° = 2 \cos \frac{\pi}{5}. \tag{1.43}$$

The formula (1.43), obtained as a result of mathematical analysis of the geometric relationships of the *decagon*, is another evidence of the fundamental nature of the *golden ratio*, which along with the number $\pi$ can rightfully be ranked as the most important mathematical constants. In the formula (1.43), the number 5 is encountered twice, and the angle 36° is the angle at the apex of the *pentagram*. Therefore, it is no accident that the number 5 was considered as *sacred number* by the Pythagoreans, and the *pentagonal star* (*pentagram*) was a symbol of the Pythagorean union!

Now, we can give interesting information from the field of atomic physics, related to decagon (Fig. 1.19). Studying the regularities of atomic nucleus, the Belarusian physicist Petrunenko in [43] concluded that the high stability of atomic nucleus is achieved due to the wave multiplicities of the *golden section*, underlying their organization. In this case, he shows that the regular decagon, based on the golden section, underlies the structure of the atomic nucleus!

## 1.8. The Golden Section in Nature

### 1.8.1. Pentagonal symmetry in Nature

In Nature, the forms, based on "pentagonal" symmetry (starfish, sea urchins, flowers), are widespread. Flowers like water-lily, hawthorn, cloves, pears, bird-cherries, apple-trees, strawberries, and many other flowers have five petals. The examples of living structures, based on "pentagonal" symmetry, are shown in Fig. 1.20.

### 1.8.2. Pentacle of Venus

In 2010, the information about the publication of the new book *Harmony: a New Way of Looking at our World* (2010) [49] was released on the Internet. The sensationalism of this information consists in the fact that the author of the book [49] is His Royal Highness Charles, The Prince of Wales, the heir to the English throne (Fig. 1.21).

The concept of "Grammar of Harmony" is introduced in the book [49]. The main idea of the book is that the Universe is the most vivid evidence of the manifestation of the "Grammar of Harmony".

In the book [49], the heir to the British throne shares his conviction that most actual problems of mankind are rooted in our disharmony with Nature, and that we can solve these problems if we will restore the balance (Harmony) with Nature.

One of the chapters of the book [49] is devoted to the elaboration of the concept of "Grammar of Harmony," which includes the main mathematical constants (like the golden proportion) and numerical sequences (like Fibonacci numbers); they permeate the whole of Nature.

Fig. 1.20. Examples of pentagonal symmetry in Nature: (a) Chinese rose; (b) an apple in the cut; (c) starfish; (d) cactus.

In his book [49], Prince Charles pays much attention to the main symbol of the Pythagoreans, the *pentacle* (Fig. 1.16), as one of the examples of the "Grammar of Harmony". Prince Charles draws attention to the "Pentacle of Venus", which is formed during the movement of Venus in the sky. As is known, every 8 years the planet Venus describes an absolutely perfect pentacle along a large circle of the celestial sphere (Fig. 1.22).

Ancient astronomers noticed this phenomenon and were so shocked that Venus and its pentacle became symbols of perfection and beauty. As if paying tribute to this phenomenon, the ancient Greeks held the Olympic Games every eight years. Today, only a few know that modern Olympic Games follow to the half cycle of Venus. Even fewer people know that the five-pointed star almost

Fig. 1.21.  His Royal Highness Charles, The Prince of Wales.

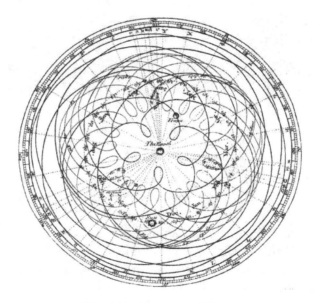

Fig. 1.22.  Pentacle of Venus.

became a symbol of the Olympic Games, but at the last moment it was modified: the five pointed ends of the star were replaced with five rings. According to organizers, such solution reflected better the goal and harmony of Olympic Games.

One could consider the wanderings of Venus across the sky, based on the pentacle and the "golden section", as some "exotic event". In modern science, many outstanding discoveries have been made; they confirm the presence of "harmonic proportions" in a huge number of natural phenomena and objects (*pentagonal symmetry in Nature*, *fullerenes* (Nobel Prize in Chemistry 1996) [116], *quasicrystals* by Dan Shechtman (Nobel Prize in Chemistry 2011) [115], *Bodnar's phyllotaxis geometry* [28], the *"golden" genomatrices* of Sergey Petukhov [48], experimental proof of the presence of harmony of the golden section in the quantum world, etc.). All this leads to the thought about the non-randomness of the Universe [49].

## 1.9. The Golden Section in Cheops Pyramid

A few centuries before Christ, there were *Seven Wonders* of the World, and many people sought to see these wonders with their own eyes. Six of these wonders, the Colossus of Rhodes and Alexandria Lighthouse on the island of Pharos, the Gardens of Babylon, the statue of Zeus at Olympia, the Temple of Artemis at Ephesus, the Mausoleum of King Karii Mavsola, were destroyed over a period of time. Only one of the "wonders of the world", the *Egyptian pyramids*, has survived.

Nikolay Vasyutinsky in his remarkable book *The Golden Proportion* describes the Egyptian pyramids very poetically [24]: "*The endless, monotonous sea of sand, rare dry bushes of plants, barely noticeable camel traces are sweeped by the wind. The hot sun of the desert ... And all here it seems dull, as if covered with fine sand. And suddenly, like a mirage, before the astonished gaze of the traveler there arise pyramids, the fantastic figures made of stone, looking to the Sun. With their enormous size, perfect geometric shape, they amaze the imagination. No wonder these creations of human hands were attributed to one of the seven wonders of the world.*"

The purpose of the pyramids (Fig. 1.23) was multifunctional. They served not only as tombs of the pharaohs, but were also attributes of the greatness, power and wealth of the country, monuments of culture, repositories of the country's history, the life

Fig. 1.23.  Complex of pyramids in Giza.

of the pharaoh and the people of this country, and the collection of household items.

It is quite clear that the pyramids had a deep "scientific content", embodied in their form, size and orientation on the terrain. Each detail of the pyramid and each element of the form had been chosen carefully and demonstrated a high standard of knowledge of the creators of the pyramids. After all, they were built on millennia, "forever". And it's not for nothing that the Arab proverb says: *"Everything in the world is afraid of time. The time is afraid of the pyramids."*

Among the grand pyramids of Egypt, a special place is occupied by the Great Pyramid of Pharaoh Cheops (Khufu). Before we begin to analyze the shape and size of the Cheops pyramid, we should recall the Egyptian system of measures. The Egyptians had three units of length: the *"elbow"* (466 mm), equal to the seven *"palms"* (66.5 mm) that, in turn, was equal to the four *"fingers"* (16.6 mm).

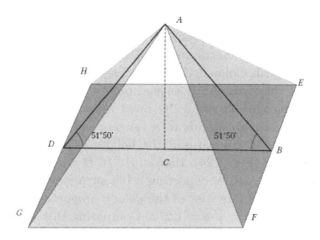

Fig. 1.24. Geometrical model of the Cheops pyramid.

Let's consider the geometric model of the Cheops pyramid (Fig. 1.24) and analyze the dimensions of the Cheops pyramid, following the arguments, given in [24].

Most researchers agree that the length of the side of the base of the pyramid, for example, $GF$ in Fig. 1.24, is equal to $L = 233.16$ m. This value corresponds almost exactly to the 500 "*elbows*". The full correspondence to the 500 "elbows" will be if the length of the "*elbow*" is considered equal to 0.4663 m.

The height of the pyramid $H$ is estimated by researchers differently from 146.6 to 148.2 m, and, depending on the adopted height of the pyramid, all the relationships of its geometric elements will change. What is the reason for the difference in the height of the pyramid? The point is that, strictly speaking, the pyramid of Cheops had been truncated. Its upper platform today has the size of about $10 \times 10$ m, and a century ago it was equal to $6 \times 6$ m. Obviously, the top of the pyramid was dismantled, and it does not meet the original one.

By estimating the height of the pyramid, it is necessary to take into consideration such a physical factor as the "sediment" of the structure. For a long time under the influence of enormous pressure (reaching 500 tons on $1\,\mathrm{m}^2$ of the lower surface), the height of the pyramid decreased in comparison with the initial height. What was the original height of the pyramid? This height can be recreated if you find the main "geometric idea" of the pyramid.

In 1837, English Colonel G. Wise measured the angle of incline of the faces of the pyramid: it turned out to be equal, $\alpha = 51°51'$. This value is recognized today by most researchers. The specified value of the angle corresponds to a tangent equal to 1.27306. This value corresponds to the ratio of the height $AC$ of the pyramid to half of its base $CB$ (Fig. 1.24), that is, $AC/CB = H/(L/2) = 2H/L$. Here, the researchers were expecting a big surprise! The fact is that if we calculate the square root of the *golden proportion* $\sqrt{\Phi}$, then we get the following result: $\sqrt{\Phi} = 1.272$. Comparing this value with the value of tg $\alpha = 1.27306$, we see that these values are very close to each other. If we take the angle $\alpha = 51°50'$, that is, we reduce it by only one corner minute, then the value of tg $\alpha$ becomes equal to 1.272, that is, it coincides with the value $\sqrt{\Phi} = 1.272$. It should be noted that in 1840, G. Wise repeated his measurements and clarified that the value of the angle $\alpha = 51°50'$.

These measurements led the researchers to the following very interesting hypothesis: the following basic relation $AC : CB = \sqrt{\Phi} = 1.272$ underlies the triangle $ACB$ of the pyramid of Cheops! Above, we found the geometric relations of such a triangle, called the *"golden" right-angular triangle* or *Kepler's triangle* (Fig. 1.14).

If we now take as a basis the hypothesis that the main "geometrical idea" of the Cheops pyramid is *Kepler's triangle* (Fig. 1.14), then, from here, we can easily calculate the "design" height of the pyramid of Cheops. It is equal to

$$\boxed{H = (L/2)/\sqrt{\Phi} = 148.28\,\text{m}.}$$

Let's derive some other geometric parameters for the Cheops pyramid, which follow from the *"golden" hypothesis*. In particular, let us calculate the ratio of the outer area of the pyramid to the area of its base. To do this, we take the length of the leg $CB$ equal to 1, that is, $CB = 1$. But then the length of the side of the base of the pyramid $GF = 2$, and the area of the base $EFGH$ will be equal to $S_{EFGH} = 4$.

Now let's calculate the area of the lateral face of the Cheops pyramid $S_\Delta$. Since for the case $CB = 1$, the height $AB$ of the triangle $AEF$ is equal to $\Phi$, then the area of the side face will be equal to $S_\Delta = \Phi$. Then the total area of all four sides of the pyramid will be equal to $4\Phi$, and the ratio of the total outer area of the pyramid to the basic area of the pyramid will be equal to the *golden proportion!* According to Nikolay Vasyutinsky [24], *this is the main geometric secret of the Cheops pyramid!*

The analysis of the other Egyptian pyramids shows that the Egyptians always sought to embody some of the important mathematical knowledge in their pyramids. In this respect, *Chephren's pyramid* is very interesting. The measurements of the pyramid showed that the angle of inclination of the side faces is equal to $\alpha = 53°12'$, which corresponds to the ratio 4:3 of the legs of a right-angular triangle. This relationship of the legs corresponds to the well-known rectangular triangle with the sides 3:4:5; this triangle is called a *"sacred"* or *"Egyptian"* *triangle*. According to historians, the Egyptians gave to the *"Egyptian"* *triangle* a magical meaning. Plutarch wrote that the Egyptians compared the Nature of the Universe with the *"sacred"* *triangle*; they symbolically likened the vertical cathetus to the husband, the foundation to the wife, and the hypotenuse to what is born from both.

For the "Egyptian" triangle 3:4:5, the following equality is true: $5^2 = 4^2 + 3^2$. This equality expresses the *Pythagoras theorem.* Possibly, the Egyptian priests wanted to immortalize the *Pythagoras' theorem* by creating the *Chephren* pyramid on the base of the *"Egyptian"* *triangle*? It is difficult to find a more successful example for illustrating the *Pythagoras theorem,* which was known to the Egyptians long before its proof by Pythagoras.

Thus, the ingenious creators of the Egyptian pyramids sought to impress their distant descendants with the depth of their geometric knowledge, and they achieved this by choosing the *"golden"* *rectangular triangle* as the "main geometric idea" for the *Cheops pyramid*, and the *"Egyptian"* *triangle* for the *Chephren pyramid*.

## 1.10. The Golden Section in Ancient Greek Culture

### 1.10.1. The magnificent Parthenon

The ancient Greeks left for us magnificent monuments of architecture that present for us the same aesthetic pleasure as to our distant ancestors. Among them, the first place rightfully belongs to the *Parthenon.*

The construction of the Parthenon is associated with the dramatic pages of the history of ancient Hellas. In 480 BC, the army of Persians, led by King Xerox, invaded Greece. The army of Persian barbarians moved from the north and stopped at the Thermopile gorge. Their path was blocked by 300 Spartan warriors, who had been covering the retreat of major troops. As a result of the betrayal, they all were killed together with their leader, King Leonid. The Persian army captured and crushed Athens.

But the Greeks with honor passed the ordeal, by defeating the Persian fleet and army. The victory of the Greeks over the Persians meant the triumph of the principles of democracy and freedom; it led to a new fruitful impulse in Greek arts, to the era of high classical art. In the works of this period, feelings of grandeur and joy dominated. Forms of art works were distinguished by increased harmony, plasticity, humanism. Parthenon, the magnificent building of the Athenian Acropolis, is the embodiment of these qualities.

During the 15 years of Pericles' reign in Athens, temples, altars, and sculptures of unusual beauty were built. The outstanding Greek sculptor Phidias was appointed the Head of all works. Recall that in Phidias' honor, the golden ratio is often called the *number* Φ.

The entire second half of the 5th century BC, the construction of temples, the altar, and the statue of Athena continued on the Acropolis. In 447 BC, the building of the temple Parthenon was started. This construction continued until 434 BC. To create a harmonious composition on the hill, its builders increased the height of the hill and erected a powerful embankment. Modern researchers have found that the length of the hill in front of the Parthenon, the length of the temple of Athens and the Acropolis section behind the Parthenon related as segments of the golden proportion.

Fig. 1.25. Parthenon.

Thus, the golden proportion had been used already for creation of the composition of temples on the sacred hill (Fig. 1.25).

Joint efforts of architects, sculptors, and the entire population of ancient Greece led to the creation of the Parthenon (Fig. 1.25). This temple, dedicated to the patroness of the city, the goddess Athena Parthenos, is rightfully considered as one of the greatest examples of ancient Greek architecture, a masterpiece of world art. It was built in the middle of the 5th century BC by sculptors Iktin and Callicrate. This was a period of the highest blossoming of ancient culture. Till date, the temple of the goddess Athena on the Acropolis hill proudly reminds about this aspect to the whole world.

Harmonic analysis of the Parthenon was carried out by many researchers. Although these studies differ somewhat in their approaches, all researchers agree with the main point: the Parthenon is remarkable for its majesty and deep humanism of architectural and sculptural images, and that the main reason for the beauty of

Fig. 1.26. Harmonic analysis of the Parthenon.

the Parthenon was the exceptional proportionality of its parts based on the golden section (Fig. 1.26).

### 1.10.2. The golden section in Greek sculpture

The great creations of Greek sculptors: Phidias, Policlet, Myron, Praxitel, in particular, the sculptures of the heroes of antiquity, for a long time and rightfully are considered as the standard of beauty of the human body, as models of a harmonious physique. In their creations, the Greek masters used the golden proportion principle. One of the highest achievements of the classical Greek art is a statue of Dorifor created by Polyclet in the 5th century BC (Fig. 1.27(a)). This statue is considered as the best example for analyzing the proportions of an ideal human body. It was this sculpture that was given the name "Canon". The figure of the young man expresses the unity of the beautiful and valiant, underlying the Greek principles of art. The statue is full of quiet confidence; harmonies of lines, balance of parts personify physical strength. Wide shoulders are almost equal to the height of the trunk, the height of the head is eight times

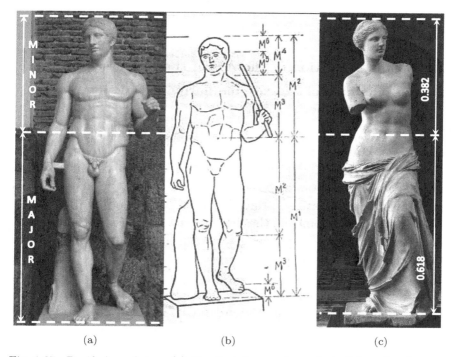

Fig. 1.27. Dorifor's sculpture (a), Dorifor's harmonic analysis (b), and sculpture of Venus of Melos (c).

inscribed into the body height, and the golden ratio corresponds to the position of the navel on the body of the athlete (see Fig. 1.27(b)). The ratio of "major" to "minor" in Fig. 1.27(a) is equal to the golden ratio.

One of the best monuments of Greek sculptural art is also Venus of Melos, the statue of the goddess Aphrodite, found on the island of Melos (sculptor Agesander from Antioch on Medar, around 120 BC). In the era of Hellenism, she was one of the favorite goddesses. She was represented as flirtatious, sometimes pensive, and sometimes playful 64t. Aphrodite from the island of Melos is strict and restrained. Probably, she stood on a high pedestal and looked at the viewer from the top down. Venus of Melos is a real pearl of the Louvre, the standard of feminine beauty of ancient Hellas. The harmonic analysis of Venus of Melos leads to a harmonic proportion that represents the "golden section".

## 1.11.  Golden Section in the Art of the Renaissance

### 1.11.1.  The idea of the "divine harmony" in the Renaissance

The idea of harmony was among the conceptual constructions of ancient culture, to which the church showed a great interest. According to the Christian doctrine, the Universe is the creation of God and unquestioningly obeys his will. The Christian God is guided by mathematical principles in the creation of the world. This Catholic doctrine in the science and art of the Renaissance has taken the form of searching for a mathematical plan, by which God created the Universe.

According to the modern American historian of mathematics, Morris Klein [101], it was precisely the close fusion of the religious doctrine about the God as the creator of the Universe and of the ancient idea of the numerical harmony of the Universe was one of the most important reasons for the great surge in culture during the Renaissance. Most clearly, the main objective of the science of the Renaissance is set forth in the following statement of the famous astronomer of this era, Johannes Kepler:

> *"The main goal of all studies of the external world should be to discover the rational order and harmony that the God has sent down to the world and revealed to us in the language of mathematics".*

The art of the Renaissance (especially painting) is largely associated with biblical subjects. A vivid example of a painting, written on biblical subjects, is Michelangelo's painting The Holy Family (Fig. 1.28).

The picture is rightly recognized as one of the masterpieces of the Western European art. The figure of Mary, Joseph and the baby Christ forms a spiral group, bringing a strong charge of plastic energy into the composite whole. Having made a harmonic analysis of this picture, the researchers found that the picture is based on the pentacle (Fig. 1.28).

Fig. 1.28. The painting "The Holy Family" by Michelangelo.

Another example of the picture, based on the biblical story, is the painting "Crucifixion" by another famous painter of the Renaissance, Raphael Santi (1483–1520). Harmonic analysis of the picture (Fig. 1.29) shows that the composition plan of this picture is based on the "golden" isosceles triangle.

### 1.11.2. "David" Michelangelo

An example of a standard model of a harmoniously developed human body is the famous statue of "David" by Michelangelo (Fig. 1.30).

In 1501, Michelangelo received the proposal from the government of Florence for the creation of the 5.5-meter statue of *David*. He worked on this statue from 1501 to 1504. This statue, installed on the main square of Florence, became a symbol of freedom of the republic. Michelangelo had depicted *David* as a beautiful, athletic giant, full of confidence and formidable power before the battle.

Fig. 1.29. The painting "Crucifixion" by Rafael Santi.

And like *Polykleitos Doryphoros*, who, in the ancient era, became the "Canon" of the beauty of the male body, David's statue can be considered as the "Canon" of the Renaissance. A comparison of David's statue (Fig. 1.30) with the statues of Doryphorus and Venus of Melos (Figs. 1.27(a)–(c)) shows that all proportions of David's statue are based on the golden section!

### 1.11.3. The painting "Mona Lisa" by Leonardo da Vinci

In the hall of the Louvre, in Paris, every visitor tries to locate one picture — the famous "Mona Lisa" or "Gioconda", belonging to the brush of Leonardo da Vinci. The master worked on his portrait for a long time. He did a lot of sketching; a lot of attention was paid to the position of the portrait, the rotation of her head, and the

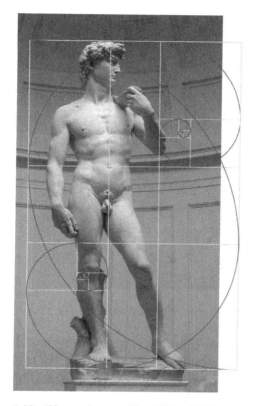

Fig. 1.30. The sculpture "David" by Michelangelo.

position of her hands. The famous artist Giorgio Vasari says that while working on the portrait of Mona Lisa, Leonardo had invited singers, musicians, and jesters to the studio to not only maintain the cheerful mood of the young female model, but also be able to follow the change in the expression of her face.

It was only after four years of hard work that he finally managed to show his famous "Gioconda" to the world. Creating his masterpiece, the artist used a secret, known to many portraitists: the vertical axis of the canvas passes through the pupil of the left eye, which should evoke a feeling of excitement in the viewer, that is, the artist used the "principle of symmetry" in his picture. But maybe the reason is different? This painting by the brilliant Leonardo attracted

Fig. 1.31. Harmonic analysis of the painting "Mona Lisa" (Gioconda).

the attention of researchers, who discovered that the compositional construction of the painting is based on the two gold isosceles triangles, turned to each other by their bases (Fig. 1.31). Thus, Leonardo used in his picture not only the principle of symmetry, but also the golden section!

What is the reason for the charm of "Mona Lisa"? It is believed that the writing of this picture was connected with some mystery in the life of Leonardo. The mystery of Leonardo begins with his birth. As we know, Leonardo was the illegally born son of the woman, about whom almost nothing is known. It is only known that her name was Katerina and that she was the mistress of the tavern. Much more is known about the father of Leonardo. The father of Leonardo, Mr. Piero, who was about 25 at the time of the birth of his son, was a notary and possessed impressive male virtues: he had lived to 77 years old, had four wives and was the father of 12 children; the last child appeared into the world, when Mr. Piero was 75 years old.

The Renaissance looked tolerantly at illegitimately born children. These children were often treated the same way as children born in wedlock. Leonardo's Father's family immediately recognized him. However, he was not immediately taken to his father's house. Shortly after his birth, he was sent along with Katerina to the village of Anchiano, located near the town of Vinci, and remained there for about 4 years; during this time, Mr. Piero married the first of his wives, a 17-year girl, who occupied a higher social status than Leonardo's mother. The young wife remained barren. Perhaps, for this reason, Leonardo, at the age of about four and half years, was taken to the city house, where he immediately found himself in the care of numerous relatives: grandfather, grandmother, father, uncles and adoptive mother. In the tax register, referring to 1457, Leonardo is called the illegal son of Piero.

Throughout his life, Leonardo kept the memory of his mother Katherine, who turned out to be surprisingly similar to Mona Lisa — the wife of the Florentine merchant Giocondo. And, perhaps, this fact was the main reason for Leonardo's desire to create a picturesque portrait of Mona Lisa. Leonardo da Vinci had not experienced such a huge tide of creative power for a long time. Everything that was in him, the brightest and the most beautiful, he invested in his work. And all his love for his mother Katherine, he embodied in his famous painting, which determined the development of painting art for many centuries to come. We thank the God for linking up Mona Lisa and Leonardo da Vinci at the sunset of the great artist's life journey!

# Chapter 2

# Fibonacci and Lucas Numbers

## 2.1. A History of the Fibonacci Numbers

### 2.1.1. Who is Fibonacci?

The cases of rampant inquisition, of the fires, on which witches and heretics were burned, and the crusades behind the "Body of the Lord" are associated with the "Middle Ages". Science in this period was clearly not "in the center of attention of society". Under these conditions, the emergence of the mathematical book *Liber Abaci* (*The Book about the Abacus*), written in 1202 by the Italian mathematician Leonardo of Pisa (known by the nickname of Fibonacci), was an important event in the "scientific life of society".

Who is Fibonacci? Why are his mathematical works so important for the history of mathematics and modern science? To answer these questions, we need to reproduce the historical era, in which Fibonacci lived and worked.

It should be noted that the period from the 11th century to the 12th century was the era of the brilliant blossoming of Arab culture, but at the same time, the beginning of its decline. By the end of the 11th century, that is, by the beginning of the Crusades, the Arabs were undoubtedly the most enlightened nation in the world, surpassing their Christian enemies in this regard. Even before the Crusades, Arab influence penetrated the West. However, the greatest penetration of Arab culture in the West began after the Crusades, which on the one hand weakened the Arab nation and on the other hand increased the Arab influence on the Christian West.

The Christian West seeks and values in the Arab world not only cotton and sugar of Palestine, pepper and ebony of Egypt, gems and spices of India. It begins to understand the cultural heritage of the "great ancient East", the custodian of which had become the Arab culture. The opened up Arab world dazzled the Christian West with its colors and scientific achievements; therefore, the demand for Arabic geographic maps, textbooks on algebra and astronomy, and Arabic architecture was becoming ever more widespread in the Western society.

One of the most interesting personalities of the era of the Crusades, the forerunner of the Renaissance, was the emperor, Friedrich Hohenstaufen, the pupil of the Sicilian Arabs and the admirer of Arab culture. At his palace in Pisa, the greatest of European mathematicians of the Middle Ages, Leonardo of Pisa (nicknamed Fibonacci, which means "son of Bonacci") lived and worked (see Fig. 2.1).

Little is known about Fibonacci's life. Even the exact date of his birth is unknown. It is assumed that Fibonacci was born in the eighth decade of the 12th century (presumably in 1170). His father was a merchant and a government official, the representative of a new class of businessmen, born during the "Commercial Revolution". At that time, Pisa was one of the largest commercial centers that actively cooperated with the Islamic East, and Fibonacci's father actively

Fig. 2.1. Leonardo of Pisa (Fibonacci).

traded in one of the trading posts, founded by the Italians on the north coast of Africa. Due to this circumstance, he managed to "find a place" for his son, the future mathematician, Fibonacci, in one of the Arab educational institutions, where Fibonacci was able to get a good mathematical education.

One of the famous historians of mathematics, Morris Cantor, called Fibonacci *a brilliant meteor that flashed against the dark background of the Western European Middle Ages.* He suggests that Fibonacci may have been killed during one of the Crusades (presumably in 1228) while accompanying Emperor Frederick Hohenstaufen.

Fibonacci wrote several mathematical works: *Liber abaci; Liber quadratorum; Practica geometriae.* The book *Liber abaci* is the most famous among them. This book was published during the life of Fibonacci in two editions in 1202 and 1228. The book consists of 15 chapters, which consistently concerns about the following: the new digital signs of Indians and how to use them to represent numbers (Chapter 1); multiplication, addition, subtraction and division of numbers (Chapters 2–5); multiplication, addition, subtraction and division of numbers with fractions (Chapters 6 and 7); finding the prices of goods and their exchange; the rule of partnership and the rule of "double false position" (Chapters 8–13); finding square and cubic roots (Chapter 14); the rules relating to geometry and algebra problems (Chapter 15).

Note that Fibonacci conceived his book as a guide for merchants, but in its significance, it went far beyond the limits of trading practice and essentially represented a kind of mathematical encyclopedia of the Middle Ages. From this point of view, Chapter 12 is of particular interest; in this chapter, Fibonacci formulated and solved a number of mathematical tasks, having an interest from the point of view of the general prospects for the development of mathematics. This chapter occupies almost a third part of the book Liber abaci and, apparently, Fibonacci gave this book the most significance and in it, Fibonacci showed the greatest originality.

Although Fibonacci was one of the most prominent mathematical minds in the history of Western European mathematics and made a huge contribution to its development, his contribution to mathematics was undeservedly belittled. Most convincingly, the Russian mathematician Prof. A.V. Vasiliev in his book *The Whole Number* (1919) noted the significance of the mathematical creativity of Fibonacci for Western mathematics:

> *"The books of the learned merchant from Pisa were so much higher of the standard of mathematical knowledge even of scientists of that time, that their influence on mathematical literature became noticeable only two centuries after his death, namely at the end of the 15th century, when many of his theorems and problems were introduced by the friend of Leonardo da Vinci, a professor at many Italian universities, Luca Pacioli in his books, and at the beginning of the 16th century, when a group of talented Italian mathematicians: Scipio del Ferro, Jerome Cardano, Tartaglia, Ferrari marked the beginning of higher algebra by solution of cubic and biquadrate equations."*

It follows from this statement that Fibonacci had outstripped almost by two centuries the Western European mathematicians of his time. Like Pythagoras, who obtained his "scientific education" from the Egyptian and Babylonian priests and then contributed to the transfer of these knowledge into Greek science, Fibonacci obtained his mathematical education from the Arab educational institutions and he tried to "implement" many of the knowledge obtained there, in particular, the Arab–Hindu decimal numeral system, in Western European science. And like Pythagoras, the historical significance of Fibonacci for the Western world was that he contributed to transfer the mathematical knowledge of Arabs into Western European science through his mathematical books and thus he laid the foundation for the further development of Western European mathematics.

Fibonacci also made an important mathematical discovery, which played a major role in the development of the mathematical theory of harmony. We are talking about the recurrent numerical sequence called *Fibonacci numbers*: $1, 1, 2, 3, 5, 8, 13, \ldots$ Each term of this sequence is equal to the sum of two previous ones. Later, it turned out that this sequence is very often found in Nature, in particular, in botany and underlies the botanical phenomenon of phyllotaxis.

## 2.1.2. The task of rabbits' reproduction

Ironically, Fibonacci, who made outstanding contributions to the development of mathematics, came to be known in modern mathematics only as the author of the interesting numerical sequence called *Fibonacci numbers*. This numerical sequence was obtained by Fibonacci when solving the task of rabbits' reproduction, and this discovery glorified him. The formulation and solution of this task is considered as the main Fibonacci contribution to the development of combinatorics. It was with the help of this task that Fibonacci anticipated the method of recurrent relations, which is considered as one of the powerful methods for solving combinatorial problems. The recurrent formula, obtained by Fibonacci in solving this task, is considered to be the first recurrent formula in the history of mathematics.

The essence of the task of rabbits' reproduction is as follows:

*"Let there be on the first day of January a pair of rabbits (female and male) in the fenced place. This pair of rabbits produces a new pair of rabbits (female and male) on the first day of February and then on the first day of each following month. Each newborn pair of rabbits becomes mature in one month and then in one month gives life to a new pair of rabbits. There is the question: how many pairs of rabbits will be in the fenced place through one year, that is, through 12 months from the beginning of reproduction?"*

To solve this task, which is shown in Fig. 2.2, we denote a pair of mature rabbits through $A$ and a pair of newborn rabbits through $B$. Then the process of "rabbits' reproduction" can be described by using two "transitions" that correspond to the monthly transformations of rabbits during the reproduction process:

$$A \to AB, \qquad (2.1a)$$

$$B \to A. \qquad (2.1b)$$

Note that the transition (2.1a) is modeling the monthly transformation of each mature pair of rabbits $A$ into two pairs $AB$, namely, into the pair of the same mature rabbits $A$ and the newborn pair of rabbits $B$. The transition (2.1b) is modeling the process of "maturing" of rabbits, when each newborn pair of rabbits $B$ during

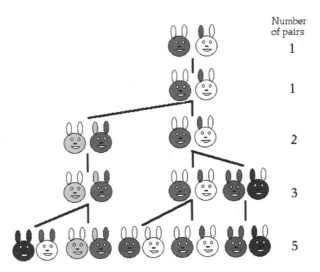

Fig. 2.2.  Fibonacci rabbits.

Table 2.1.

| Data | Couple of rabbits | $A$ | $B$ | $A + B$ |
|------|-------------------|-----|-----|---------|
| 1 January | $A$ | 1 | 0 | 1 |
| 1 February | $AB$ | 1 | 1 | 2 |
| 1 March | $ABA$ | 2 | 1 | 3 |
| 1 April | $ABAAB$ | 3 | 2 | 5 |
| 1 May | $ABAABABA$ | 5 | 3 | 8 |
| 1 June | $ABAABABAABAAB$ | 8 | 5 | 13 |

one month turns into the mature pair $A$. Then, if we start in the first month from the mature couple $A$, then the process of rabbits' reproduction can be demonstrated by using Table 2.1.

Note that columns $A$ and $B$ of Table 2.1 show the number of mature and newborn pairs of rabbits in each month of the year and column $A + B$ shows the total number of rabbits. By studying the number of rabbits in $A$, $B$ and $(A + B)$ columns of Table 2.1, we can find the following regularity: the number of rabbits in the $A$, $B$ and $(A + B)$ columns of Table 2.1 is equal to the sum of rabbits in two previous columns. If we now denote the number of rabbits in $i$th row of Table 2.1 through $F_i$, then the above general regularity can be written in the following mathematical form:

$$F_i = F_{i-1} + F_{i-2}. \tag{2.2}$$

Such a formula is called a *recurrent formula* (from the Latin word *recurer*, meaning return).

Note that the specific values of the numerical sequence, generated by the recurrent formula (2.2), depend on the initial values $F_1$ and $F_2$ of the sequence. For example, we have $F_1 = F_2 = 1$ for $A$-numbers, and for this case, the recurrent formula (2.2) "generates" the following numerical sequence:

$$1,\ 1,\ 2,\ 3,\ 5,\ 8,\ 13,\ 21,\ 34,\ 55,\ 89,\ 144,\ 233,\ \ldots \tag{2.3}$$

For the $B$-numbers, we have $F_1 = 0$ and $F_2 = 1$; then the corresponding numerical sequence for this case will be

$$0,\ 1,\ 1,\ 2,\ 3,\ 5,\ 8,\ 13,\ 21,\ 34,\ 55,\ 89,\ 144,\ 233,\ \ldots$$

Finally, for $(A + B)$ numbers, we have $F_1 = 1$ and $F_2 = 2$; then the corresponding numerical sequence for this case will be

$$1,\ 2,\ 3,\ 5,\ 8,\ 13,\ 21,\ 34,\ 55,\ 89,\ 144,\ 233,\ \ldots$$

In mathematics, the numerical sequence (2.3) is usually named *Fibonacci numbers*.

### 2.1.3. Variations on Fibonacci theme

The sequence of the Fibonacci numbers has a number of amazing mathematical properties. Many famous mathematicians of the 20th century were fascinated by the Fibonacci numbers. Here, first of all, it is necessary to mention the studies of the Canadian geometer Harold Coxeter (1907–2003) [7] and the Russian mathematician Nikolay Vorobyov (1925–1995), who in 1961 published a remarkable brochure, *Fibonacci Numbers* [8], which had been translated into many world languages.

One of the mathematicians, who had been interested in the Fibonacci numbers, was the famous Hungarian mathematician Alfred Rényi (1921–1970). He made significant contributions to the development of probability theory, mathematical statistics, information theory, combinatorics, graph theory, number theory and mathematical analysis. A deep musician and literature expert, Rényi highly evaluated the aesthetic principle in mathematics and was also interested in the history of mathematics and reflected on its philosophical problems.

Rényi was a brilliant popularizer of mathematics. And this can be seen while reading the collection of his scientifically popular articles, *The Trilogy on Mathematics* [18]. One of the articles "Variations on the Fibonacci Theme" is devoted to the Fibonacci numbers. In this article, exploring the properties of the Fibonacci numbers, he compares his researches on the Fibonacci numbers with musical variations on a given topic, a genre well known in musical literature.

Mozart, Beethoven and other composers were great lovers of this genre. A distinctive feature of the works of the variational genre is the fact that in most cases, variations begin with one simple main theme, which undergoes further significant changes in tempo, mood and character. But no matter how bizarre the variations change, there arise the impression in the listeners that all variations are a natural development of the main theme.

By using this musical genre as a basis, Rényi selects a very simple mathematical topic (the Fibonacci numbers) and further develops this topic along with numerous variations. These variations are different in their properties and allow different interpretations; they have different applications and have varying degrees of generality.

## 2.2. The Sums of the Consecutive Fibonacci Numbers

**The sum of the $n$ consecutive Fibonacci numbers:** Let's start with the simplest sums of this kind:

$$
\begin{aligned}
1 + 1 &= 2 = \mathbf{3} - 1, \\
1 + 1 + 2 &= 4 = \mathbf{5} - 1, \\
1 + 1 + 2 + 3 &= 7 = \mathbf{8} - 1, \\
1 + 1 + 2 + 3 + 5 &= 12 = \mathbf{13} - 1.
\end{aligned}
\tag{2.4}
$$

If you pay attention to the numbers in bold $3, 5, 8, 13$, then you can easily find that they are a sequence of the Fibonacci numbers. From this observation, we can write the general formula for the sum of the $n$ consecutive Fibonacci numbers:

$$F_1 + F_2 + \cdots + F_n = F_{n+2} - 1. \tag{2.5}$$

**The sum of the $n$ consecutive Fibonacci numbers with odd indices:** To do this, we start with the simplest examples:

$$
\begin{aligned}
1 + 2 &= \mathbf{3}, \\
1 + 2 + 5 &= \mathbf{8}, \\
1 + 2 + 5 + 13 &= \mathbf{21}, \\
1 + 2 + 5 + 13 + 34 &= \mathbf{55}.
\end{aligned}
\tag{2.6}
$$

The analysis of (2.6) allows us to find the following regularity: the sum of the $n$ consecutive Fibonacci numbers with the *odd* indices is always equal to the Fibonacci number with the even index. In general, the partial sums (2.6) can be represented by the following mathematical formula:

$$F_1 + F_3 + F_5 + \cdots + F_{2n-1} = F_{2n}. \tag{2.7}$$

**The sum of the $n$ consecutive Fibonacci numbers with even indices:** It is easy to prove a similar formula for the sum of the Fibonacci numbers with the *even* indices:

$$F_2 + F_4 + F_6 + \cdots + F_{2n} = F_{2n+1} - 1. \tag{2.8}$$

**The sum of squares of the consecutive Fibonacci numbers:** Now, we find what is equal to the following sum:

$$F_1^2 + F_2^2 + \cdots + F_n^2. \tag{2.9}$$

As always, let's start with the analysis of the simplest sums of the kind (2.9):

$$
\begin{aligned}
1^2 + 1^2 &= 2 = \mathbf{1 \times 2}, \\
1^2 + 1^2 + 2^2 &= 6 = \mathbf{2 \times 3}, \\
1^2 + 1^2 + 2^2 + 3^2 &= 15 = \mathbf{3 \times 5}, \\
1^2 + 1^2 + 2^2 + 3^2 + 5^2 &= 40 = \mathbf{5 \times 8}.
\end{aligned}
\tag{2.10}
$$

Analysis (2.10) leads us to the following general mathematical formula:

$$F_1^2 + F_2^2 + \cdots + F_n^2 = F_n F_{n+1}, \tag{2.11}$$

that is, the sum of squares of the consecutive Fibonacci numbers is equal to the product of the largest Fibonacci number in this sum and the next Fibonacci number after it.

**The sum of the squares of the two adjacent Fibonacci numbers:** Let's find the value of the following sum:

$$F_{n-1}^2 + F_n^2. \tag{2.12}$$

Let's start by calculating the simplest sums of the kind (2.12):

$$
\begin{aligned}
1^2 + 1^2 &= 1 + 1 = \mathbf{2}, \\
1^2 + 2^2 &= 1 + 4 = \mathbf{5}, \\
2^2 + 3^2 &= 4 + 9 = \mathbf{13}, \\
3^2 + 5^2 &= 9 + 25 = \mathbf{34}.
\end{aligned}
\tag{2.13}
$$

The analysis of (2.13) allows us to find another interesting regularity: the sum of the squares of the two adjacent Fibonacci numbers $F_{n-1}^2$ and $F_n^2$ is always equal to the Fibonacci number $F_{2n-1}$:

$$F_{n-1}^2 + F_n^2 = F_{2n-1}. \tag{2.14}$$

We will accept without proof [11] a number of remarkable properties of the Fibonacci numbers:

$$F_m F_n + F_{m-1} F_{n-1} = F_{m+n-1}, \tag{2.15}$$

$$F_{n+1} F_m + F_n F_{m-1} = F_{m+n}. \tag{2.16}$$

In particular, for $m = n$, from (2.15) follows the identity (2.14) and from (2.16) follows the identity (2.17):

$$F_{2n} = (F_{n-1} + F_{n+1})F_n = (2F_{n-1} + F_n)F_n. \tag{2.17}$$

By enjoying and admiring the formulas (2.5), (2.7)–(2.9), (2.11)–(2.17) in depth, one can understand the delight of many prominent mathematicians of the 20th century, in particular, the Russian mathematician Nikolay Vorobyov, the author of the brochure, *Fibonacci Numbers* [8], which is considered to be the mathematical bestseller of the 20th century and the American mathematician Verner Hoggatt, the creator of the American Fibonacci Association, the founder of the mathematical journal *The Fibonacci Quarterly* and the author of Ref. [9]. It was in the Fibonacci numbers, associated with the *golden ratio*, that they saw some *mathematical secrets of Nature*, and this inspired them to devote their mathematical talent to the exploration of this unique mathematical phenomenon.

## 2.3. Cassini's Formula

### 2.3.1. Who is Giovanni Cassini?

Cassini is a famous dynasty of the French astronomers. The founder of this dynasty, Giovanni Domenico Cassini (1625–1712) is considered to be the most famous of them. Many astronomical objects are named after Giovanni Cassini: *Cassini's Crater* on the Moon, *Cassini's Crater* on the Mars, Cassini's Laws are the three astronomical laws discovered by Giovanni Cassini (see Fig. 2.3).

*Cassini–Huygens* is a spacecraft, created jointly by NASA, the European Space Agency and the Italian Space Agency, whose goal is to study planet Saturn, and its rings and satellites. The device consisted of two main components: the *Cassini Orbiter station* itself and the Huygens descent probe, which was separated from the station and descended onto the surface of the satellite *Saturn's Titan*. *Cassini–Huygens* was launched on October 15, 1997 and reached Saturn on July 1, 2004. It is the first artificial satellite on Saturn.

### 2.3.2. Cassini's contribution to the development of the theory of Fibonacci numbers (Cassini's formula)

It turns out that, Cassini's name is widely known not only in astronomy but also in mathematics. The history of science is silent on, why Cassini was fascinated by the Fibonacci numbers. Most likely, it was just a "hobby" of the great astronomer. At that time, many serious scientists studied the Fibonacci numbers and the golden

Fig. 2.3.  Giovanni Domenico Cassini.

section. Recall that these mathematical objects were also the passion of Johannes Kepler, a contemporary of Cassini.

Cassini first drew attention to the following regularity, linking the neighboring Fibonacci numbers. If we take the arbitrary Fibonacci number, for example, $F_5 = 5$ and square it, we get the following result: $5^2 = 25$. And now let's compare this result with the product of two adjacent Fibonacci numbers $F_4 = 3$ and $F_6 = 8$, which surround the Fibonacci number $F_5 = 5$, that is, $3 \times 8 = 24$. We find that the compared numbers differ by 1, that is,

$$5^2 - 3 \times 8 = 1.$$

Let's do the same with the next *triple* of the following Fibonacci numbers $5, 8, 13$, that is, first we raise the Fibonacci number $F_6 = 8$ in a square ($8^2 = 64$), then we compare this result with the product of the two adjacent to $F_6 = 8$ Fibonacci numbers 5 and 13 ($5 \times 13 = 65$). To our surprise, we find that the compared numbers also differ by 1, that is,

$$8^2 - 5 \times 13 = -1.$$

In this case, however, the resulting difference is also equal to 1, taken with the negative sign.

Next, we have $13^2 - 8 \times 21 = 1$, $21^2 - 13 \times 34 = -1$, and so on.

As a result of these elementary considerations, Cassini discovered a surprising regularity that can be formulated as follows:

> *"The square of some Fibonacci number $F_n$ is always different from the product of the two adjacent Fibonacci numbers $F_{n-1}$ and $F_{n+1}$ by 1. The sign of this 1 depends on the index n of the Fibonacci number $F_n$; if the index n is an **even** number, then the number 1 is taken with a **minus**, and if **odd**, then with a **plus**."*

This property of the Fibonacci numbers can be expressed by the following general mathematical formula called *Cassini formula*:

$$F_n^2 - F_{n-1}F_{n+1} = (-1)^{n+1}, \tag{2.18}$$

which is true for the integer number $n = 0, \pm 1, \pm 2, \pm 3, \ldots$.

In conclusion, we note the aesthetic character of Cassini's formula (2.18) as well as other identities for the Fibonacci numbers (2.4)–(2.17). The formula (2.18) is awe-inspiring if we imagine that it is valid for the arbitrary value of the index $n$ (in the following, we show that the index $n$ can be the integer number from $-\infty$ to $+\infty$), and in true aesthetic sense, because the alternation $+1$ and $-1$ in the above mathematical formula (2.18) at the sequential consideration of all Fibonacci numbers from $-\infty$ to $+\infty$ causes an unconscious sense of rhythm and harmony.

## 2.4. Lucas Numbers

### 2.4.1. Who is Lucas?

Fibonacci did not continue to study the mathematical properties of the Fibonacci numbers. Other mathematicians did this instead of him. Starting from the 19th century, the mathematical works on the properties of the Fibonacci numbers, according to the witty remark of a mathematician, *began to multiply like Fibonacci rabbits*. The

Fig. 2.4. Lucas Francois Edouard Anatole (1842–1891).

French mathematician, Lucas, became the leader of these studies in the 19th century (see Fig. 2.4).

Lucas Francois Edouard Anatole was born in 1842 and died in 1891 as a result of an accident that occurred at a banquet, when fragments of a broken dish injured his cheek. Lucas died from infection a few days later.

Lucas' most important mathematical works relate to number theory and indefinite analysis. In 1878, Lucas found a criterion for determining the fact whether a *Mersenne number* $M_p = 2^p - 1$ is simple or composite. By using his method, Lucas found that *Mersenne's number* $M_{127} = 2^{127} - 1 = 170141183460469231731687303715884105727$ is the prime number. For 75 years, this number remained the largest *prime* number known in mathematics.

Let's give some explanations for these Lucas scientific results. It is well known that the *prime* numbers are such numbers that do not have other divisors besides themselves and unit 1: $2, 3, 5, 7, 11, 13, \ldots$. It was already known to the Pythagoreans that there are infinitely many *prime numbers* (the proof of this statement is in Euclid's *Elements*). The study of the *prime numbers* and the determination of their distribution in the series of natural numbers is a very difficult task of the theory of numbers. Therefore, the scientific result, obtained by Lucas in the field of the *prime numbers*, undoubtedly, belongs to the category of the outstanding mathematical achievements.

It is curious that Lucas studied the so-called *perfect numbers*. It is proved that as we move from the beginning in the natural series, the perfect numbers meet less and less. In the first 10 000 numbers of the natural series, there are only four *perfect numbers*. The search for the *perfect numbers* turned out to be a fascinating activity for mathematicians. The fifth *perfect number* $2^{12}(2^{13} - 1)$ was found in the 15th century by the German mathematician, Regiomontanus. In the 16th century, the German scientist Sheibel found the two *perfect numbers*, 8 589 869 056 and 137 438 691 328.

In 1644, the French mathematician, Mersenne, found the eighth *perfect number* $2^{30}(2^{31} - 1)$. Lucas in the 19th century found the *12th perfect number*. Research in this area continues to this day, and here all powers of modern computers are used. For example, the 18th perfect number, which was found by using computer simulation, has 2000 decimal digits.

In Lucas's honor, it is necessary to note one more of his scientific achievements. Lucas already in the 19th century, that is, long before the advent of modern computers, drew attention to the technical advantages of the binary system for the implementation of computing devices and machines, that is, he had outstripped John von Neumann, the eminent American physicist and mathematician, who gave a strong preference to the binary system for the technical implementation of electronic computers (*John von Neumann Principles*).

From the point of view of "harmonious" mathematics, Lucas's most important scientific achievement was the introduction of the concept of the *Fibonacci numbers* as well as the introduction of the concept of the *Lucas numbers* that are calculated by the same recurrent relation as the classical Fibonacci numbers, but for other initial conditions.

## 2.4.2. Generalized Fibonacci numbers

Lucas introduced the so-called *generalized Fibonacci numbers*, described by the following recurrent formula:

$$G_n = G_{n-1} + G_{n-2}. \tag{2.19}$$

In dependence of the initial terms $G_1, G_2$, the recurrent formula (2.19) generates an infinite number of numerical sequences similar to the classical Fibonacci numbers.

Among all the possible sequences, calculated by the recurrent relation (2.19), the two numerical sequences, the *Fibonacci numbers* (2.3) and the so-called *Lucas numbers*, are the most known; the Lucas numbers are calculated by the following recurrent formula:

$$L_n = L_{n-1} + L_{n-2} \qquad (2.20)$$

at the following initial terms:

$$L_1 = 1 \quad \text{and} \quad L_2 = 3. \qquad (2.21)$$

Then, by using the recurrent formula (2.20) and the initial conditions (2.21), we can calculate the recurrent numerical sequence called *Lucas numbers*:

$$1, 3, 4, 7, 11, 18, 29, 47, 76, 123, 199, \ldots. \qquad (2.22)$$

For the *Lucas numbers* (2.22), we can obtain many remarkable properties, similar to the properties of the Fibonacci numbers, derived above [8, 9, 11]. We consider only some of them:

$$L_1 + L_2 + \cdots + L_n = L_{n+2} - 3,$$

$$L_1 + L_3 + L_5 + \cdots + L_{2n-1} = L_{2n} - 2,$$

$$L_2 + L_4 + L_6 + \cdots + L_{2n} = L_{2n+1} - 1,$$

$$L_1^2 + L_2^2 + \cdots + L_n^2 = L_n L_{n+1} - 2,$$

$$L_n^2 + L_{n+1}^2 = 5F_{2n+1},$$

$$\lim_{n \to \infty} \frac{L_n}{L_{n-1}} = \Phi = \frac{1 + \sqrt{5}}{2}.$$

### 2.4.3.  The extended Fibonacci and Lucas numbers

Until now, we have considered Fibonacci and Lucas numbers, implying that their indices $n$ are natural numbers $n = 1, 2, 3, \ldots$. It turns out that they can be extended to negative values of the

Table 2.2. The extended Fibonacci and Lucas numbers.

| $n$ | 0 | 1 | 2 | 3 | 4 | 5 | 6 | 7 | 8 | 9 | 10 |
|-----|---|---|---|---|---|---|---|---|---|---|----|
| $F_n$ | 0 | 1 | 1 | 2 | **3** | 5 | **8** | 13 | **21** | 34 | **55** |
| $F_{-n}$ | 0 | 1 | −1 | 2 | **−3** | 5 | **−8** | 13 | **−21** | 34 | **−55** |
| $L_n$ | 2 | 1 | 3 | 4 | 7 | **11** | 18 | **29** | 47 | **76** | 123 |
| $L_{-n}$ | 2 | **−1** | 3 | **−4** | 7 | **−11** | 18 | **−29** | 47 | **−76** | 123 |

indices $n$, that is, when the indices $n$ take the values from the set: $n = 0, -1, -2, -3, \ldots$. At the same time, new unusual properties of these sequences are found.

The extended Fibonacci and Lucas numbers are presented in Table 2.2. As follows from Table 2.2, the terms of the extended sequences $F_n$ and $L_n$ have a number of remarkable mathematical properties. For example, for the *odd* indices $n = 2k + 1$, the terms of the extended Fibonacci numbers $F_n$ and $F_{-n}$ are the same, that is, $F_{2k+1} = F_{-2k-1}$ and for the *even* $n = 2k$, they are opposite in sign, that is, $F_{2k} = -F_{-2k}$ (see the Fibonacci numbers in bold in Table 2.2). As for the extended Lucas numbers $L_n$, we have opposite results, that is, $L_{2k} = L_{-2k}$ and $L_{2k+1} = -L_{-2k-1}$ (see the Lucas numbers in bold in Table 2.2).

And now let's compare the extended Fibonacci and Lucas sequences given by Table 2.2. Let's consider, for example, the Lucas number $L_4 = 7$ and compare it with the extended Fibonacci sequence. It is not difficult to find that $L_4 = 7 = 5 + 2$. But 2 and 5 are the extended Fibonacci numbers $F_3 = 2$ and $F_5 = 5$.

But maybe our observation is a random coincidence? By continuing the study of Table 2.2, we get that $1 = 0 + 1$, $3 = 1 + 2$, $4 = 1 + 3$, $7 = 2 + 5$, $11 = 3 + 8$, $18 = 5 + 13$ and so on *ad infinitum*.

Let's compare numerical sequences $L_{-n}$ and $F_{-n}$. Here, we find the same, that is, $-1 = 0 + (-1)$, $3 = 1 + 2$, $-4 = (-1) + (-3)$ and so on. Thus, we have established the following surprisingly simple mathematical relation, connecting the extended Fibonacci and Lucas numbers:

$$L_n = F_{n-1} + F_{n+1}, \qquad (2.23)$$

where $n$ take the following values: $n = 0, \pm 1, \pm 2, \pm 3, \ldots$.

By continuing the study of Table 2.2, it can also find that the extended Fibonacci and Lucas numbers are linked by the very simple mathematical relations, for example:

$$L_n = F_n + 2F_{n-1}, \tag{2.24}$$

$$L_n + F_n = 2F_{n+1}. \tag{2.25}$$

## 2.5. Binet's Formulas

### 2.5.1. Who is Binet?

The French mathematician Jacques Phillip Marie Binet (1786–1856) became another Fibonacci enthusiast of the 19th century. He was known as a well-known French mathematician and astronomer and was also, a member of the Paris Academy of Sciences (Fig. 2.5).

Binet was born on February 2, 1776 in Rennes (France). In 1804, he entered the Polytechnic School in Paris and after graduating in 1806, he worked in the department of bridges and roads of the French government. In 1807, he became a teacher at the Polytechnic School, and one year later, became an Assistant Professor of applied analysis and descriptive geometry.

Fig. 2.5.  Jacques Phillip Marie Binet (1786–1856).

He had published many articles on mechanics, mathematics, and astronomy. In mathematics, Binet introduced the term *beta-function*; he studied linear difference equations with variable coefficients, etc. Binet investigated the foundations of the theory of matrices, and his work in this direction was continued by other researchers. In 1812, he discovered the *rule of matrix multiplication*, and this discovery already glorified his name more than the other studies. In 1843, Binet was elected to the Paris Academy of Sciences.

However, Binet entered into the theory of Fibonacci numbers as the author of the famous mathematical formulas, known in mathematics under the name of *Binet's formulas*. These formulas connect the Fibonacci and Lucas numbers with the golden proportion and, undoubtedly, belong to the category of the outstanding mathematical formulas.

The researches of Lucas and Binet became the launching pad, in the second half of the 20th century, for the victorious march toward the Fibonacci Association, organized by the group of American mathematicians in 1963.

## 2.5.2. Deducing Binet's formulas

Two centuries after the scientific discoveries of Johannes Kepler, who more than some of his contemporaries, appreciated the role of the *golden section* in the future development of science and compared it with the *Pythagoras theorem*, in the 19th century, the interest in the Fibonacci numbers, the Lucas numbers, and the golden ratio in mathematics started to increase.

In this regard, we should mention about the two enthusiasts of the Fibonacci numbers, the French mathematicians, Lucas and Binet. Above, we talked about the Lucas numbers (2.22), which the French mathematician Francois Edouard Anatol Lucas (1842–1891) introduced into consideration. In essence, Lucas, with his mathematical research, revived the interest of the scientific community of the 19th century in the Fibonacci numbers and the golden section.

In this section, we discuss the remarkable *Binet formulas*, introduced into consideration by the French mathematician Binet

(1786–1856). In order to derive *Binet's formulas*, Binet used the roots $x_1 = \frac{1+\sqrt{5}}{2}$ and $x_2 = \frac{1-\sqrt{5}}{2}$ of the golden section equation $x^2 - x - 1 = 0$. By using these roots, we can obtain the analytical expressions for the extended Fibonacci and Lucas numbers.

Let us represent the extended Fibonacci numbers $F_n$ ($n = 0, \pm 1, \pm 2, \pm 3, \ldots$) as follows:

$$F_n = k_1 x_1^n - k_2 x_2^n. \tag{2.26}$$

The constant coefficients $k_1, k_2$ in (2.26) are easily calculated if we use the initial values for the Fibonacci numbers $F_0 = 0$, $F_1 = 1$. By using (2.26), we construct the following system of algebraic equations:

$$F_0 = k_1 - k_2 = 0,$$
$$F_1 = k_1 x_1 - k_2 x_2 = 1. \tag{2.27}$$

The solution of the system of the equations (2.27) leads to the following results:

$$k_1 = k_2, \tag{2.28}$$

$$k_1 x_1 - k_2 x_2 = k_1(x_1 - x_2) = k_1 \sqrt{5} = 1. \tag{2.29}$$

From (2.28) and (2.29), it follows that

$$k_1 = k_2 = \frac{1}{\sqrt{5}}. \tag{2.30}$$

By substituting (2.30) into (2.26), we obtain the Binet formula for the Fibonacci numbers:

$$F_n = \frac{\left(\frac{1+\sqrt{5}}{2}\right)^n - \left(\frac{1-\sqrt{5}}{2}\right)^n}{\sqrt{5}}. \tag{2.31}$$

Let's represent the extended Lucas numbers $L_n$ ($n = 0, \pm 1, \pm 2, \pm 3, \ldots$) as follows:

$$L_n = x_1^n + x_2^n \tag{2.32}$$

or

$$L_n = \left(\frac{1 + \sqrt{5}}{2}\right)^n + \left(\frac{1 - \sqrt{5}}{2}\right)^n. \qquad (2.33)$$

For the cases $n = 0$ and $n = 1$, we get the following initial values of the extended Lucas numbers:

$$L_0 = 2; \quad L_1 = 1. \qquad (2.34)$$

To be fair, it should be noted that the formulas (2.31) and (2.33) were derived by Abraham de Moivre (1667–1754) and Nikolay Bernoulli (1687–1759) one century before Jacques Binet. However, in the modern mathematical literature, the formulas (2.31) and (2.33) are called *Binet's formulas*.

### 2.5.3. Proof of Cassini formula

Binet's formulas are widely used to prove various Fibonacci identities. As an example, we consider the proof of the *Cassini formula* (2.18), based on the *Binet formula* (2.31). To do this, we represent the formula (2.31) as follows:

$$F_n = \frac{\Phi^n - \left(-\frac{1}{\Phi}\right)^n}{\sqrt{5}}. \qquad (2.35)$$

By substituting (2.35) into the expression (2.18), we get the following:

$$F_n^2 - F_{n+1}F_{n-1} = \left(\frac{\Phi^n - \left(-\frac{1}{\Phi}\right)^n}{\sqrt{5}}\right)^2$$

$$- \left(\frac{\Phi^{n+1} - \left(-\frac{1}{\Phi}\right)^{n+1}}{\sqrt{5}}\right)\left(\frac{\Phi^{n-1} - \left(-\frac{1}{\Phi}\right)^{n-1}}{\sqrt{5}}\right)$$

$$= \left(\frac{\Phi^{2n} - 2(-1)^n + \left(-\frac{1}{\Phi}\right)^{2n}}{5}\right)$$

$$- \left(\frac{\Phi^{2n} - (-1)^{n-1}\Phi^2 - (-1)^{n-1}\left(\frac{1}{\Phi}\right)^2 + \left(-\frac{1}{\Phi}\right)^{2n}}{5}\right)$$

$$= \frac{1}{5}\left(-2(-1)^n - (-1)^n\left(\Phi^2 + \Phi^{-2}\right)\right)$$

$$= \frac{1}{5}\left(-2(-1)^n - 3(-1)^n\right)$$

$$= -(-1)^n = (-1)^{n+1}. \qquad (2.36)$$

Note that for simplification in (2.36), we used the following elementary identity, based on the Binet formula for Lucas numbers given by (2.33):

$$L_2 = \Phi^2 + \Phi^{-2} = \left(\frac{1+\sqrt{5}}{2}\right)^2 + \left(\frac{1-\sqrt{5}}{2}\right)^2$$

$$= \frac{1 + 2\sqrt{5} + 5 + 1 - 2\sqrt{5} + 5}{2} = 3.$$

## 2.6. Steinhaus's "Iron Table"

The famous Polish mathematician Hugo Steinhaus, who is considered to be an internationally recognized expert in the field of probability theory, built a table of random numbers by using the golden proportion. For this purpose, he multiplied 10 000 integers from 1 to 10 000 by the number $\phi = \Phi - 1 = 0.61803398$, where $\Phi$ is the golden proportion. As a result of multiplication, he obtained a sequence of numbers, multiplied by $\phi$:

$$1\phi, \; 2\phi, \; 3\phi, \ldots, 4181\phi, \ldots, 6765\phi, \ldots, 10\,000\phi.$$

Steinhaus called this numeral sequence the *"golden numbers"*. Each "golden number" contains integer and fractional parts. For example, the number $1000\phi = 618.03398$ has the integer part 618 and the fractional part 0.03398; the number $4181\phi = 2584.0001$. Moreover, he established that there are no *"golden numbers"* with the fractional part, equal to 0, and also there are no two *"golden numbers"* with equal fractional parts. Thus, each "golden number" has a single fractional part.

If we now arrange all the *"golden numbers"* in accordance with their fractional parts, we will see that the smallest fractional part will have the number 4181 and the largest the number 6765. If we now arrange 10 000 natural numbers in increasing order of their fractional parts, we obtain the following table of natural numbers:

| 4181 | 8362 | 1597 | 5778 | 9959 |
|------|------|------|------|------|
| 3194 | 7365 | 0610 | 4791 | 8972 |
| $\vdots$ | $\vdots$ | $\vdots$ | $\vdots$ | $\vdots$ |
| 8739 | 1974 | 6155 | 3571 | 7752 |
| 0987 | 5168 | 9349 | 2584 | 6765 |

Steinhaus called the resulting table *Iron Table,* taking into consideration a number of its unique mathematical properties. The *Iron Table* demonstrates the deep links with the Fibonacci numbers. The first property is that the difference between adjacent numbers of the *Iron Table* is always equal to one of the numbers 4181, 6765 and 2584. Indeed, we have

$$8362 - 4181 = 4181, \quad 8362 - 1597 = 6765,$$

$$5778 - 1597 = 4181, \ldots, \quad 9349 - 5168 = 4181,$$

$$9349 - 2584 = 6765, \quad 6765 - 2584 = 4181.$$

We can very easily determine the numbers 2584, 4181 and 6765 if we consider the Fibonacci series:

$$1, 1, 2, 3, 5, 8, 13, 21, 34, 55, 89, 144, 233, 377, 610, 897,$$

$$1597, \mathbf{2584}, \mathbf{4181}, \mathbf{6765}, \ldots.$$

Thus, the characteristic numbers of the *Steinhaus Table* 2584, 4181, 6765 are nothing but the three neighboring Fibonacci numbers:

$$F_{18} = 2584, \quad F_{19} = 4181, \quad F_{20} = 6765.$$

We can see from the *Iron Table* that it starts with the Fibonacci number $F_{18} = 2584$ and ends with two adjacent Fibonacci numbers $F_{19} = 4181$, $F_{20} = 6765$.

It is clear that the *Iron Table* can be constructed for an arbitrary number $N$ of natural numbers. Jan Grzędzielski in Ref. [20] analyzed the *Iron Tables* for the cases $N = F_n$, where $F_n$ is the Fibonacci number. At the same time, he discovered an interesting regularity that occurs when we move from the *Iron Table* with $N = F_{n-1}$ to the next *Iron Table* with $N = F_n$. The *Iron Table*, corresponding to $N = F_n$, as it were, "moves apart" in comparison with the previous *Iron Table*, corresponding to $N = F_{n-1}$, creating strictly defined positions in the new *Iron Table* for numbers $F_{n-1} + 1, F_{n-1} + 2, \ldots, F_n - 1, F_n$.

According to Grzędzielski [20], the design method of the *Iron Table* "reminds the functioning of all radiation spectra in Nature".

## 2.7. Pythagorean Triangles and Their Presentation Through Fibonacci and Lucas Numbers

### 2.7.1. Pythagoras theorem

As is known, the *Pythagoras theorem* is perhaps the most famous theorem of geometry. The essence of this theorem is extremely simple. Let's consider the right-angular triangle with the sides $\{a, b, c\}$, where $a$ is the hypotenuse and $b$ and $c$ are the legs of the right-angular triangle. The *Pythagoras theorem* states that in the right-angular triangle, the legs $b$ and $c$ are connected with the hypotenuse $a$ by the following simple relation:

$$a^2 = b^2 + c^2. \tag{2.37}$$

Despite its utmost simplicity, the *Pythagoras theorem*, according to many mathematicians, belongs to the category of the most famous mathematical theorems in the history of mathematics.

### 2.7.2. Pythagorean triangles

Among the infinite number of right-angular triangles, satisfying the relation (2.36), the so-called *Pythagorean triangles*, whose sides are integers, have been always of particular interest. The right-angular triangle with the sides $\{3, 4, 5\}$ is the most widely known (Fig. 2.6). It was also called *sacred* or *Egyptian*, as it was widely used in Egyptian culture. As mentioned above, it is this triangle that represents the main geometric idea of the *Khefren Pyramid* in Giza.

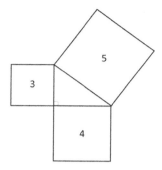

Fig. 2.6. "Sacred" or "Egyptian" triangle.

For the Egyptian triangle in Fig. 2.6, the *Pythagoras theorem* (2.37) looks extremely simple:

$$3^2 + 4^2 = 5^2. \tag{2.38}$$

There is a legend that the relation (2.38) was used by Egyptian land surveyors and builders to determine the right angle on a plane. For this, a rope was used, for example, 12 m long, which was divided into three parts with lengths 3, 4 and 5 m by special loops or knots. To determine the right angle, the Egyptian surveyor stretched one part of the rope, for example, 3 m long, and fixed it on the ground with the help of special "pegs", hammered into two loops. Then the rope was stretched with the help of the third loop and this loop was fixed with the help of a "peg". It is clear that the angle, formed between the two smaller sides of the triangle, was exactly equal to 90°. It was believed that during the laying of the pyramids, such a ritual procedure for determining the right angles of the basis of the pyramid on the ground was performed by Pharaoh himself.

### 2.7.3. Fibonacci's "Pythagorean triangles"

The following question arises: Are there any Pythàgorean triangles of a kind different from the Egyptian triangle (see Fig. 2.6)? The surprising answer to this question is given in [108].

Let's consider the "four" neighboring Fibonacci numbers

$$F_n, \ F_{n+1}, \ F_{n+2}, \ F_{n+3}. \tag{2.39}$$

As an example of (2.39), we consider the following "four" neighboring Fibonacci numbers:

$$1, 2, 3, 5. \tag{2.40}$$

Let's now consider the following procedure, which will lead us to an infinite number of Fibonacci's "Pythagorean triangles":

(1) Multiply the two middle or the internal Fibonacci numbers from (2.40): $2 \times 3 = 6$. For the general case (2.39), we must calculate the product: $F_{n+1} \times F_{n+2}$.

(2) Double the result: $2 \times 6 = 12$. For the general case (2.39), we must calculate the product $c = 2 \times (F_{n+1} \times F_{n+2})$. The resulting number $c$ is equal to the first side (leg) of the desired Pythagorean triangle.

(3) Now, we multiply the two external Fibonacci numbers from (2.40): $1 \times 5 = 5$. For the general case (2.39), we must calculate the product: $b = F_n \times F_{n+3}$. The number $b$ represents the second side (leg) of the Pythagorean triangle.

(4) The third, the longest side (hypotenuse), is found by summing the squares of the internal numbers from (2.40): $2^2 = 4$ and $3^2 = 9$, that is, their sum is equal: $4 + 9 = 13$. For the general case (2.39), we have $a = F_{n+1}^2 + F_{n+2}^2$.

It is easy to verify that the sides of the obtained right-angular triangle really form a *Pythagorean triangle* because $12^2 + 5^2 = 13^2$.

For the general case (2.39), the sides of the Pythagorean triangle are connected by the following relation:

$$(2 \times F_{n+1} \times F_{n+2})^2 + (F_n \times F_{n+3})^2 = (F_{n+1}^2 + F_{n+2}^2)^2. \tag{2.41}$$

By direct calculations, it is easy to verify that the identity (2.41) holds for all possible "fours" of the Fibonacci numbers of the kind (2.39). Indeed, for the case $n = 1$, the "four" Fibonacci numbers (2.39) take the form:

$$1, 1, 2, 3. \tag{2.42}$$

In accordance with the above algorithm, we can calculate the sides of the Pythagorean triangle for this case:

$$c = 2 \times 1 \times 2; \quad b = 1 \times 3 = 3; \quad a = 1^2 + 2^2 = 1 + 4 = 5.$$

Thus, the simplest case (2.42) corresponds to the *sacred* or *Egyptian* triangle, for which the *Pythagoras theorem* takes the form (2.38).

Let's consider the *Pythagorean triangle* for the case $n = 3$. In this case, the "four" of the neighboring Fibonacci numbers look as follows:

$$2, 3, 5, 8. \tag{2.43}$$

Then, in accordance with the above algorithm, the sides of the *Pythagorean triangle* are as follows:

$$c = 2 \times 3 \times 5 = 30; \quad b = 2 \times 8 = 16; \quad a = 3^2 + 5^2 = 9 + 25 = 34.$$

The *Pythagoras theorem* for this case looks as follows:

$$30^2 + 16^2 = 34^2.$$

Finally, for the case $n = 4$, the "four" neighboring Fibonacci numbers look as follows:

$$3, 5, 8, 13, \tag{2.44}$$

and the sides of the Pythagorean triangle are respectively equal to

$$c = 2 \times 5 \times 8 = 80; \quad b = 13 \times 3 = 39; \quad a = 5^2 + 8^2 = 89.$$

The *Pythagoras theorem* looks as follows:

$$80^2 + 39^2 = 89^2.$$

Table 2.3 gives an idea of the Fibonacci's "Pythagorean triangles" for the initial values of $n = 1, 2, 3, \ldots, 8$.

It is essential to emphasize that the side $a$ of the Pythagorean triangles from Table 2.3 is calculated by the following formula:

$$a = F_{n+1}^2 + F_{n+2}^2. \tag{2.45}$$

Table 2.3. Fibonacci's "Pythagorean triangle".

| $n$ | $F_n$ | $F_{n+1}$ | $F_{n+2}$ | $F_{n+3}$ | $c$ | $b$ | $a$ |
|---|---|---|---|---|---|---|---|
| 1 | 1 | 1 | 2 | 3 | 4 | 3 | 5 |
| 2 | 1 | 2 | 3 | 5 | 12 | 5 | 13 |
| 3 | 2 | 3 | 5 | 8 | 30 | 16 | 34 |
| 4 | 3 | 5 | 8 | 13 | 80 | 39 | 89 |
| 5 | 5 | 8 | 13 | 21 | 205 | 105 | 233 |
| 6 | 8 | 13 | 21 | 34 | 546 | 272 | 610 |
| 7 | 13 | 21 | 34 | 55 | 1428 | 715 | 1597 |
| 8 | 21 | 34 | 55 | 89 | 3740 | 1869 | 4181 |

By using the identity (2.14), we can write down $a = F_{2n+3}$, that is, the hypotenuse $a$ of the Fibonacci's "Pythagorean triangle" is always equal to the certain Fibonacci number that is confirmed by Table 2.3.

### 2.7.4. Lucas' "Pythagorean triangles"

It turns out that the above procedure for constructing *Fibonacci's Pythagorean triangles* is also valid for the Lucas numbers (2.22). For example, the first "four" of the neighboring Lucas numbers 1, 3, 4, 7 lead us to *Lucas's Pythagorean triangle* with the sides:

$$c = 2 \times 3 \times 4 = 24; \quad b = 1 \times 7 = 7; \quad a = 3^2 + 4^2 = 9 + 16 = 25.$$

For this *Lucas Pythagorean triangle*, the *Pythagoras theorem* looks as follows:

$$24^2 + 7^2 = 25^2.$$

In Table 2.4, we construct the different *Lucas Pythagorean triangles* for the initial values of $n = 1, 2, 3, \ldots, 8$.

The second "four" 3, 4, 7, 11 of the neighboring Lucas numbers (2.22) lead us to another *Lucas Pythagorean triangle* with the sides:

$$c = 2 \times 4 \times 7 = 56; \quad b = 3 \times 11 = 33; \quad a = 4^2 + 7^2 = 16 + 49 = 65.$$

Table 2.4. Lucas's "Pythagorean triangles".

| $n$ | $L_n$ | $L_{n+1}$ | $L_{n+2}$ | $L_{n+3}$ | $c$ | $b$ | $a$ |
|---|---|---|---|---|---|---|---|
| 1 | 1 | 3 | 4 | 7 | 24 | 7 | 25 |
| 2 | 3 | 4 | 7 | 11 | 56 | 33 | 65 |
| 3 | 4 | 7 | 11 | 18 | 154 | 72 | 170 |
| 4 | 7 | 11 | 18 | 29 | 396 | 203 | 445 |
| 5 | 11 | 18 | 29 | 47 | 1044 | 517 | 1165 |
| 6 | 18 | 29 | 47 | 76 | 2726 | 1368 | 3050 |
| 7 | 29 | 47 | 76 | 123 | 7144 | 3567 | 7985 |
| 8 | 47 | 76 | 123 | 199 | 18696 | 9353 | 2095 |

For this *Lucas Pythagorean triangle*, the *Pythagoras theorem* has the following form:

$$56^2 + 33^2 = 65^2.$$

The above connection of the Fibonacci and Lucas numbers with the *Pythagoras theorem* proves the existence of an infinite number of the Fibonacci and Lucas Pythagorean triangles, which is additional evidence of the fundamental importance of the Fibonacci and Lucas numbers for geometry.

## 2.8. Fibonacci Numbers in Nature

### 2.8.1. Fibonacci rectangles

To build "Fibonacci rectangles" (Fig. 2.7), we use the Fibonacci sequence $1, 1, 2, 3, 5, 8, 13, 21, \ldots$.

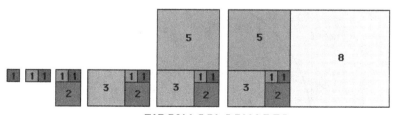

FIBONACCI SQUARES

Fig. 2.7. Fibonacci rectangles.

Let's start with a square with the side 1. Let's take two such squares (with the area of 1) and connect them together. As a result, we get the Fibonacci rectangle of the dimension 2×1. Such a rectangle is called *double square*. Then, on the larger side of the "double square," we build the new square of the dimension 2 × 2. As a result, we get the Fibonacci rectangle of the dimension 3 × 2. On the larger side of this rectangle, we build a new square with the dimension 3×3; as a result, we get the new Fibonacci rectangle of the dimension 5×3. By continuing this process, we will successively get the Fibonacci rectangles (Fig. 2.7), in which the sides are the adjacent Fibonacci numbers, that is, they have dimensions 8 × 5, 13 × 8, 21 × 13 and so on.

### 2.8.2. Fibonacci spirals

Now in each of the squares, which form the Fibonacci rectangles, we draw the arc that represents a quarter of a circle. By combining these arcs, we get the curve, which resembles a spiral by its shape (Fig. 2.8). Strictly speaking, this curve is not a spiral from a mathematical point of view, but it is a very good approximation to spirals, which are widely found in Nature. In the future, we will name the curve in Fig. 2.8 as the *Fibonacci spiral.*

The great poet and naturalist Goethe considered spirality as one of the characteristic features of all living organisms, the manifestation of the most intimate essence of life. The antennae of plants and ram's horns are twisted spirally; the sunflower seeds are arranged spirally

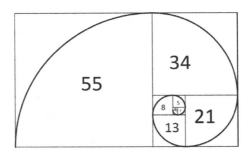

Fig. 2.8.  Fibonacci spiral.

Fig. 2.9. The nautilus spiral.

in a sunflower head. Each of us would have many times admired the shape of seashells, which are also built according to a spiral law. But, after all, our Galaxy also has a spiral shape! The shape of the nautilus shell (Fig. 2.9) is an example of the use of the Fibonacci spiral.

### 2.8.3. Phenomenon of phyllotaxis and Fibonacci numbers

As is known, the Fibonacci and Lucas numbers underlie the botanical "law of phyllotaxis" [28]. According to this law, the number of left and right spirals on the surface of the so-called phyllotaxis objects (pine cone, pineapple, cactus, sunflower head, etc.) is described by the relations of the neighboring Fibonacci numbers:

$$\frac{F_{n+1}}{F_n} : \quad \frac{2}{1}, \frac{3}{2}, \frac{5}{3}, \frac{8}{5}, \frac{13}{8}, \frac{21}{13}, \cdots \quad (2.46)$$

These relationships characterize the "symmetry" of phyllotaxis objects. Moreover, each phyllotaxis object is characterized by its own

(a)                    (b)                    (c)

(d)                    (e)                    (f)

Fig. 2.10. Phyllotaxis structures: (a) cactus; (b) a sunflower head; (c) echinacea; (d) cauliflower; (e) pineapple; (f) pine cone.

relation of the neighboring Fibonacci numbers, which is called the *symmetry order* [28].

Figure 2.10 shows the examples of such objects (cactus, sunflower head, echinacea, cauliflower head, pineapple, pine cone), in which the phyllotaxis law is expressed by its symmetry order, formed by the ratios of the neighboring Fibonacci numbers, that is, in each of these botanical objects, the seeds or small parts of objects on their surface are located at the intersection of the left and right spirals; the ratio of the number of left and right spirals is always equal to one of the ratios of the neighboring Fibonacci numbers (2.46).

Figure 2.11 shows geometric models of phyllotaxis structures, which give a figurative representation of this unique botanical phenomenon.

Thus, we find strict mathematics in the arrangement of leaves on the stems of plants, petals on a rose flower, in a spiral

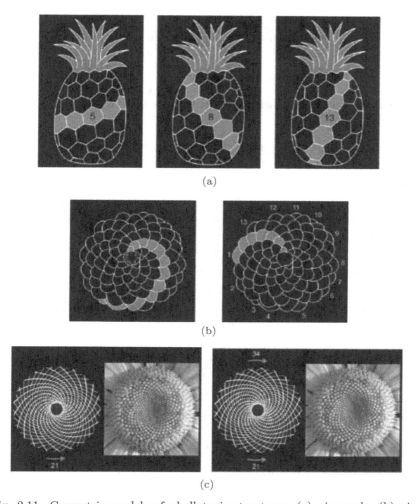

Fig. 2.11. Geometric models of phyllotaxis structures: (a) pineapple; (b) pine cone; (c) daisy.

arrangement of the seeds in a pine cone, a sunflower head, pineapple and a cactus. And this pattern is mathematically expressed by the Fibonacci numbers and the golden ratio! And again, we are convinced that everything in Nature is subjected to a single plan, a single law — "the law of the golden section"; discovering and explaining this fundamental law of Nature in all its manifestations is the main task of science.

## 2.9. Fibonacci Numbers and Solution of Hilbert 10th Problem

In the summer of 1900, mathematicians gathered at their second International Congress in Paris. The famous German mathematician, Professor of the University of Göttingen, David Hilbert (1862–1943) was invited to give one of the keynote lectures. As the world's greatest mathematician, he became famous for his work on algebra, number theory, and shortly before the Congress, he resolutely rebuilt the axioms of Euclidean geometry in his fundamental work *The Foundations of Geometry* (1899). After much hesitation, Hilbert chose the unusual form of his speech. In his lecture *Mathematical Problems*, he decided to formulate the mathematical problems, which, in his opinion, should determine the development of mathematics in the coming century.

Hilbert's appeal to the International Mathematical Congress, held in 1900 in Paris, is perhaps the most significant lecture in the history of mathematics given by the prominent mathematician for other mathematicians and devoted to the fundamental problems of mathematics. In his lecture, Hilbert outlined 23 major mathematical problems that must be solved in the new century.

Hilbert's lecture was more than a simple collection of the mathematical problems. This lecture reflected Hilbert's philosophy of mathematics and proposed the mathematical problems, which are very important from the point of view of his philosophy. And although more than a century had passed, Hilbert's lecture preserves its significance and can be read with great interest by someone interested in mathematic history and research.

As it is known, Hilbert's 10th problem is called the *Problem of Solving Diophantine Equations* and in order to explain the essence of this problem, we must turn 17 centuries back to the ancient mathematician, Diophantus.

We know very little about Diophantus, who is considered as the last great mathematician of antiquity. His work played such a significant role in the history of algebra that many historians of mathematics put a lot of effort into determining the span of his life. It is assumed that Hilbert lived in the middle of the 3rd century AD and he lived 84 years. The main work of Diophantus was "Arithmetic". It was this fundamental mathematical work, consisting of the 13 books, that was a turning point in the development of algebra and the theory of numbers. Diophantus set the task of finding integer values of algebraic equations. Such equations are called Diophantine.

In his famous 1900 lecture, David Hilbert described the 10th problem as follows:

> *"There is given the Diophantine equation with a certain number of unknowns and rational integer coefficients. It is necessary to find up the procedure, which could determine in the finite number of operations: whether the equation is solvable in rational integers."*

Hilbert's 10th Problem was solved by the young Russian mathematician, Yuri Matiyasevich. His name became widely known in 1970, when he completed the last missing step in the "negative solution" to Hilbert's 10th problem.

Yuri Matiyasevich (born on March 2, 1947, Leningrad) is the Soviet and Russian mathematician, the researcher of the St. Petersburg branch of V. A. Steklov's Mathematical Institute of the Russian Academy of Sciences (RAS), Doctor of Physics and Mathematics, Academician of the RAS. He made a significant contribution to the theory of computability through the solution to Hilbert's 10th Problem (see Fig. 2.12).

And now, we approach the answer of the main question from the point of view of the *Mathematics of Harmony*: How did Matiyasevich use the *Fibonacci numbers* for solving Hilbert's 10th problem. In one of his works, Matiyasevich wrote thus:

> *"It is well known that each natural number can be uniquely represented as the sum of the different Fibonacci numbers, in which there are no two adjacent Fibonacci numbers (the so-called Zeckendorf representation). Thus, we can consider the natural numbers as the binary words with the additional condition that in such binary words there are no two units*

Fig. 2.12.  Yuri Matiyasevich.

*next to each other. I contrived to show that with such representation of numbers as the binary words, both the sequence of words and equations, equal to the length of the two words can be expressed by the Diophantine equations."*

And further, Yuri Matiyasevich writes about the significance of the new mathematical theorem in the field of the theory of the *Fibonacci numbers*, proved by the prominent Russian mathematician Nikolay Vorobyov for the solution to Hilbert's 10th Problem:

*"... During the summer of 1969, I had read with great interest the third expanded edition of the popular book on Fibonacci numbers, written by N.N. Vorobyov from Leningrad. It seems incredible that in the 20th century it is possible to find something new about the numbers, introduced by Fibonacci else in the 13th century in connection to rabbits' reproduction.*

*However, the new edition of Vorobyov's book contained, in addition to the traditional material, some of the original results of its author. In fact, Vorobyov received them a quarter of a century earlier, but he never published them. His results caught my attention right away, but I was not yet able to use them directly to construct Diophantine representations of the exponential type...*

*Who could say what would happen if Vorobyov did include this theorem*
*in the first edition of his book? It is possible that Hilbert's 10th problem*
*would solved ten years earlier!"*

By making Yuri Matiyasevich's question more precise, we have
a right to ask the following question: What would have happened if
the Italian mathematician Fibonacci hadn't introduced the Fibonacci
numbers in the 13th century? Perhaps, Hilbert's 10th problem would
not have been solved until now. Of course, Vorobyov's theorem, used
by Yuri Matiyasevich, was the important mathematical result, but
the main "culprit" in the solution to Hilbert's 10th problem was the
Italian mathematician Leonardo from Pisa (Fibonacci). As early as
1202, he published the book *Liber Abaci*, in which he introduced the
Fibonacci numbers.

The main conclusion from these arguments is that the solution
to one of the most complex mathematical problems, Hilbert's 10th
Problem, was obtained by using the Fibonacci number theory [8, 9,
11]. And this fact in itself lifts both the theory of Fibonacci numbers
[8, 9, 11] and the "Mathematics of Harmony" [6] to a high standard.

And although Yuri Matiyasevich did not make such a contribution
to the development of the Fibonacci number theory, as Nikolay
Vorobyov or Verner Hoggatt, he was the first modern mathematician,
who showed a brilliant example of using the Fibonacci numbers to solve
one of the most complicated mathematical problems, the *Hilbert's 10th*
*problem.*

The main Matiyasevich theorem (complete solution) on
insolubility in general form of Hilbert's 10th Problem is as follows:

*"For general case, there is impossible one and the same general method*
*(algorithm), which allows to establish for the arbitrary Diophantine*
*equation: whether it has a solution in integers or not".*

## 2.10. Turing and Fibonacci Numbers

Alan Mathison Turing (1912–1954) is an English mathematician,
logician, cryptographer, inventor of the Turing machine. Alan Turing
has a wide and ever-growing reputation as one of the most creative
thinkers of the 20th century (see Fig. 2.13). His research interests

Fig. 2.13. Alan Turing (1912–1954).

(from theoretical informatics to artificial intelligence and biology) cover many new research topics of the 21st century. To mark the outstanding achievements of Alan Turing in the field of theoretical informatics, the Turing Award, the most prestigious award in computer science, given by the Association of Computing Techniques for the outstanding scientific and technical contributions in this field, was established.

Turing studied at the Sherborne School, where he showed his outstanding abilities in mathematics and chemistry, and then at the Royal College of Cambridge University, from which he graduated in 1934. His immediate teacher, and later a colleague, was the famous mathematician Newman (1897–1984); Turing listened to Newman's course on the foundations of mathematics in 1935. In the same year, Turing received the Royal College scholarship to work on his dissertation. It was during this period that Turing published the key article, outlining the scientific discovery now called the *Turing machine*, the fundamental idea in the field of computer science and mathematical logic.

During 1936–1938, Turing was an intern at the Princeton University (US), where the American logician A. Church (1903–1995) was his research advisor. After receiving his doctorate, Turing rejected John von Neumann's proposal to remain in the United

States and returned to Cambridge, where he received a Royal College scholarship to study logic and number theory, while attending L. Wittgenstein's seminars on the philosophy of mathematics. During the Second World War, Turing worked at Bletchley Park, the British cryptographic center, where he headed one of the research groups involved in decoding Enigma-coded messages from the Kriegsmarine and Luftwaffe encryption machine. By reading coded German messages, he concluded in March 1943 that, Great Britain was on the verge of defeat in the Battle of the Atlantic and in the Second World War. It is likely that without the decoding of the Enigma code, the outcome of this War would have been different.

After John von Neumann proposed a plan to create an EDVAC computer in the United States, similar work was launched in the UK at the National Physical Laboratory, where Turing worked from 1945 to 1948. The scientist proposed the highly ambitious project of Automatic Computing Engine (ACE). However, it was never implemented.

It is less widely known that in addition to research in the field of theoretical computer science, Turing was fascinated by the solution to another scientific problem, which, at first glance, has no direct relation to computer science. We are talking about *phyllotaxis*, a well-known botanical phenomenon that underlies the processes of shaping many botanical objects.

Researchers of Turing creativity assert that even at school, Turing got acquainted with the classic book *On Growth and Form,* published in 1917 by English biologist and mathematician, D'Arcy Wentworth Thompson (1860–1948). In this book, the phenomenon of phyllotaxis is considered in detail. The main mystery of phyllotaxis is how in the process of growth of a phyllotaxis object, the *Fibonacci helixes* are formed on its surface.

When Turing returned from the USA to Cambridge (1947–1948), he attended lectures in physiology and it was then that he made the first attempts to give a logical description of the nervous system and continued his studies on phyllotaxis. The first Turing article on this topic [124] was published in 1952. Unfortunately, Turing's unexpected death in 1954 did not allow him to fully uncover the mystery of phyllotaxis, related to the Fibonacci numbers. However, research in this area continues. As an example, one can cite the books on this problem, written by the Ukrainian architect Oleg Bodnar [42].

In conclusion, we can formulate the main scientific problems, the solution to which brought Alan Turing the fame of the creator of theoretical computer science and the prominent thinker of the 20th century:

(1) The Turing machine, which is an extension of the finite state machine and, according to Church–Turing's thesis, is able to imitate any abstract computer that implements the step-by-step calculation process.

(2) The decryption of the cryptographic code of the German encryption machine *Enigma* significantly influenced on the course of the Second World War.

(3) Project of Automatic Computing Machine ACE (Automatic Computing Engine).

(4) Research on phyllotaxis [124], of course, can be regarded as Turing's "hobby", which has no direct relation to its main results in the field of theoretical computer science. However, it should not be forgotten that Alan Turing was one of the eminent thinkers of the 20th century. And Turing's research on phyllotaxis, linked with his research on the creation of the logical model of the brain, the unique natural computing machine, can only be regarded as a brilliant premonition of using the Fibonacci mathematics for the creation of the Fibonacci computers.

## 2.11. Role of the Fibonacci Numbers Theory in Modern Mathematics

### 2.11.1. Fibonacci association

Studies by Lucas and Binet stimulated further research in this area of modern mathematics. In 1963, a group of US mathematicians created the Fibonacci Association. In the same year, the *Fibonacci Association* began the publication of *The Fibonacci Quarterly*. In 1984, the *Fibonacci Association* began holding regular international conferences focusing upon *Fibonacci Numbers and*

Fig. 2.14. Verner Emil Hoggatt (1921–1981).

*Source*: Verner Emil Hoggatt Jr. (1921–1981), number theorist http://faculty.
evansville.edu/ck6/bstud/hoggatt.html.

*Their Applications.* The Fibonacci Association played a significant role in stimulating future international research.

The American mathematician Verner Emil Hoggatt (1921–1981) (see Fig. 2.14), Professor at San Jose State University, was one of the founders of the *Fibonacci Association* and the journal *The Fibonacci Quarterly*.

In 1969, Hoggatt published the book, *Fibonacci and Lucas Numbers* [9], which until now is considered as one of the best books in the field. Hoggatt made timely contributions to the promotion of research in the field of the Fibonacci numbers. A scientific supervisor for many master's theses, Hoggatt authored numerous articles on the Fibonacci numbers.

The learned monk Brother Alfred Brousseau (1907–1988) (see Fig. 2.15) was another prominent founder of the *Fibonacci Association* and *The Fibonacci Quarterly*.

The main purpose of the *Fibonacci Association* was studying, first of all, the Fibonacci numbers. This fact raises some questions:

(1) Why did the members of the Fibonacci Association and many "mathematics lovers" focus so much on the Fibonacci numbers?

(2) What united these two very different persons, mathematician Verner Hoggatt and learned monk Brother Alfred Brousseau, in their quest to create the *Fibonacci Association* and establish *The Fibonacci Quarterly*?

Fig. 2.15.  Brother Alfred Brousseau (1907–1988).

In the attempt to answer these questions regarding Hoggatt's and Brousseau's narrow focus on the Fibonacci numbers, we need to examine some of their documents, in particular, their photographs, as well as books and publications.

In 1969, *TIME* magazine published Brousseau's article titled "*The Fibonacci Numbers*", which was dedicated to the *Fibonacci Association*. This article contained a photo of Alfred Brousseau with the cactus in his hands. The cactus is of course one of the most characteristic examples of the Fibonacci botanical objects. The article also referred to other natural forms involving the Fibonacci numbers, for example, the Fibonacci numbers are found in the spiral formations of sunflowers, pine cones, branching patterns of trees, and leaf arrangement (or phyllotaxis) on the branches of trees.

Alfred Brousseau recommended to the devotees of the Fibonacci numbers to *pay attention to the search for aesthetic satisfaction in them. There is some kind of mystical connection between these numbers and the Universe.*

However, Verner Hoggatt holds the pine cone in his hands in the photo (Fig. 2.14). The pine cone, of course, is another well-known example of the Fibonacci botanical objects found in Nature. From this comparison, it may be reasonable to assume that Verner Hoggatt, like Alfred Brousseau, believed in the mystical connection between the Fibonacci numbers and the Universe. *In our opinion, this belief was the main factor and primary motivation in Hoggatt and Brousseau's study of the Fibonacci numbers.*

As indicated previously, the Fibonacci numbers are associated with the *golden ratio* because the ratio of the two adjacent Fibonacci numbers strives to attain the *golden ratio* in the limit. This means that the Fibonacci numbers, approximating the *golden ratio*, express the harmony of the Universe, i.e., *there is some kind of mystical connection between these numbers and the Universe* (Alfred Brousseau).

This means that the theory of the Fibonacci numbers and their connection to the golden ratio, which began to develop more rapidly after the creation of the Fibonacci Association (1963), was primarily aimed at searching Harmony in natural objects studied in physics, chemistry, botany, biology, physiology, medicine and also in economics, computer science, education and fine arts.

Furthermore, in analyzing the origin and development of the Fibonacci number theory in modern mathematics, we are drawn back to the ancient Greek *Doctrine of the Numerical Harmony of the Universe.* Here lie the very roots of the harmonization of mathematics and theoretical natural sciences, based upon the *golden ratio* and the Fibonacci numbers [8, 9, 11, 21–23]. This approach to the "Mathematics of Harmony" [6], which is a generalization of the theory of the Fibonacci numbers [8, 9, 11, 21–23], emphasizes the interdisciplinary nature of this mathematical discipline.

### 2.11.2. The role of Nikolay Vorobyov in the development of Fibonacci number theory

In addition to the importance of the creation of the American Fibonacci Association and *The Fibonacci Quarterly,* it should be noted that the Soviet mathematician Nikolay Vorobyov (1925–1995) was the first modern mathematician to draw attention to the Fibonacci number theory (see Fig. 2.16).

In 1961, that is, long before the first US mathematical book *Fibonacci and Lucas Numbers*, published by Verner Hoggatt in 1969 [9], Vorobyov published the book *Fibonacci Numbers* [8], which also

Fig. 2.16.  Nikolay Vorobyov (1925–1995).

played a prominent role in the development of the Fibonacci numbers theory. This book became the 20th century bestseller. Apart from the publication of several editions, this book was also translated into multiple languages and became a handbook for the Soviet and international researchers on the Fibonacci numbers theory and its applications in modern theoretical natural sciences.

# Chapter 3

# Pascal Triangle, Fibonacci $p$-Numbers and Golden $p$-Proportions

### 3.1. Binomial Theorem

In combinatorics [110], the following formula, called *Newton's binomial*, is widely known:

$$(a + b)^n = C_n^0 a^n + C_n^1 a^{n-1} b + \cdots + C_n^k a^{n-k} b^k + \cdots + C_n^n b^n, \quad (3.1)$$

where $C_n^k$ represent *binomial coefficients*.

The coefficients $C_n^k$ can be calculated by using the following formula, which uses the so-called *factorials* $n! = 1 \times 2 \times 3 \times \cdots \times n$:

$$C_n^k = \frac{n!}{k!(n-k)!}. \quad (3.2)$$

From school algebra, we know the following formulas for *Newton's binomials* of the second and third degrees:

$$(a + b)^2 = a^2 + 2ab + b^2, \quad (3.3)$$

$$(a + b)^3 = a^3 + 3a^2 b + 3ab^2 + b^2. \quad (3.4)$$

Note that the name *Newton's binomial* contains historical injustice because this formula was known long before Newton to many scientists from different countries, including Al-Kashi, Tartaglia, Fermat, Pascal. Newton's achievement is as follows: He extended the formula (3.1) to some real numbers $a$ and $b$, that is, he showed that the formula (3.1) is also true when $a$ and $b$ are rational or irrational, positive or negative.

The formula (3.1) allows us to establish the following properties of the *binomial coefficients*. By taking $a = 1$ and $b = x$ in *Newton's binomial* (3.1), we obtain the following well-known function of $x$:

$$(1 + x)^n = C_n^0 + C_n^1 x + \cdots + C_n^k x^k + \cdots + C_n^n x^n. \qquad (3.5)$$

By using the formula (3.5), we can prove many properties of the binomial coefficients $C_n^k$.

Let's prove, for example, that

$$C_{n+1}^k = C_n^k + C_n^{k-1}. \qquad (3.6)$$

To do this, it suffices to multiply both sides of (3.5) by $(1 + x)$:

$$(1 + x)^{n+1} = (C_n^0 + C_n^1 x + \cdots + C_n^k x^k + \cdots + C_n^n x^n)(1 + x). \qquad (3.7)$$

The expression on the left-hand side of (3.7) can be decomposed according to (3.5). Let's replace $n$ in the formula (3.5) by $n + 1$. Therefore, the coefficient at $x^k$ will be $C_{n+1}^k$. In the right part of the formula (3.7), after opening the brackets, the element, containing $x^k$, will appear twice: when multiplying $C_n^k x^k$ by 1 and when multiplying $C_n^{k-1} x^{k-1}$ by $x$. Therefore, the coefficient at $x^k$ on the right-hand side of the equality (3.6) has the form: $C_n^k + C_n^{k-1}$; this implies the validity of the identity (3.6).

If we now accept $x = 1$ in the formula (3.5), then we obtain the following well-known identity:

$$2^n = C_n^0 + C_n^1 + \cdots + C_n^k + \cdots + C_n^n. \qquad (3.8)$$

If we take $x = -1$ in the formula (3.5), we get

$$0 = C_n^0 - C_n^1 + C_n^2 - C_n^3 + \cdots + (-1)^k C_n^k + \cdots + (-1)^n C_n^n. \qquad (3.9)$$

## 3.2. Pascal Triangle

To facilitate the calculation of binomial coefficients $C_n^k$, $k = 0, 1, \ldots, n$ in *Newton's binomial* (3.1), the great French mathematician and physicist Blaise Pascal (1623–1662), 350 years ago, suggested a

Fig. 3.1. Blaise Pascal (1623–1662).

special method, for determining the binomial coefficients, known as the *Pascal triangle* (Fig. 3.1).

To construct *Pascal's triangle*, he used the following properties of the binomial coefficients:

$$C_n^0 = C_n^n = 1; \quad C_n^k = C_n^{n-k}, \tag{3.10}$$

$$C_{n+1}^k = C_n^{k-1} + C_n^k. \tag{3.11}$$

The property (3.11) is sometimes called *Pascal's rule*. By using this rule, Blaise Pascal proposed an original way to calculate the binomial coefficients by arranging them in the form of the triangular table in Fig. 3.2. If you draw the numerical table in Fig. 3.2, you get an isosceles triangle, called *Pascal's triangle*. In this triangle, we see units on the top and sides. Each "internal" element of Pascal's triangle is equal to the sum of the two elements above it. The rows of *Pascal's triangle* are placed symmetrically about the vertical axis.

At the top of *Pascal's triangle* (Fig. 3.2), there is the single binomial coefficient $C_0^0 = 1$. This is the *zero row* of *Pascal's triangle*. The next row, called the *first row*, consists of the two units, symmetrically located relative to the unit of the zero row. These

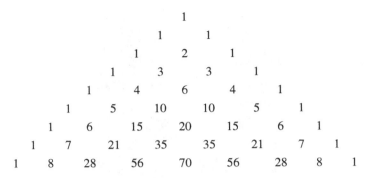

Fig. 3.2. Pascal's triangle.

are the binomial coefficients, $C_1^0 = 1$ and $C_1^1 = 1$. Each subsequent row consists of the two units located at its edges (these are binomial coefficients of the types $C_n^0 = 1$ and $C_n^n = 1$); each "internal" number of this row is formed from the two numbers of the previous row, standing above this number to the left and to the right relative to this number, according to *Pascal's rule* (3.11).

Analyzing Fig. 3.2, it is easy to formulate the following properties of *Pascal's triangle*:

(1) The sum of the numbers of the $n$th row of *Pascal's triangle* is equal to $2^n$, which corresponds to the formula (3.8).

(2) All the rows of *Pascal's triangle* are symmetrical, relative to the vertical axis, which follows from the property $C_n^k = C_n^{n-k}$.

Note that the above *Pascal's triangle* (Fig. 3.2) first appeared in Pascal's book, *A Treatise on the Arithmetic Triangle* (1665).

The binomial coefficients and *Pascal's triangle* are widely used in various branches of mathematics, informatics and other sciences. Essentially, *it is one of the fundamental and most aesthetic mathematical objects underlying the exact sciences.*

American mathematician, writer and popularizer of science Martin Gardner (1914–2010) praised Pascal's triangle as follows [12]:

> *"The Pascal triangle is so simple that even a ten-year-old child can write it out. At the same time, it keeps inexhaustible treasures and connects together various aspects of mathematics, which at first glance*

*have nothing in common with each other. Such unusual properties make Pascal's triangle one of the most elegant schemes in all mathematics. "*

## 3.3. Diagonal Sums of Pascal's Triangle and Fibonacci *p*-Numbers

### 3.3.1. Mathematical discovery of George Polya

Now, it is time to discuss another "secret" of Pascal's triangle and its connection to the Fibonacci numbers. This secret was revealed in the second half of the 20th century by several mathematicians independently. It is believed that George Polya (1887–1985), the famous Hungarian, Swiss and American mathematician and popularizer of science, made this mathematical discovery the first in the 20th century [111]. In 1940, Georse Polya immigrated to the US, where he worked a lot with school teachers of mathematics and made a great contribution to the popularization of mathematics. He had written several books about how people solve mathematical problems and how to learn to solve these problems (Fig. 3.3).

In the book *Mathematical discovery* [111], in one of the exercises, Polya described a method of obtaining the so-called "diagonal sums" of Pascal's triangle which led him to the Fibonacci numbers (Fig. 3.4).

It should be noted that this extremely simple mathematical result, which, as they say, "lay on the surface," for several centuries

Fig. 3.3. George Polya (1887–1985).

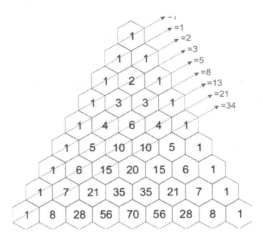

Fig. 3.4.  Diagonal sums of Pascal's triangle.

and remained as the "secret" for Blaise Pascal and other mathematicians who studied Pascal's triangle and Fibonacci numbers.

In his book [111], George Polya also proposed several tasks in the form of exercises, related to Pascal's triangle. The first task is to express the Fibonacci numbers through binomial coefficients, that is, to find the general formula for the "diagonal sums" of the Pascal triangle, as shown in Fig. 3.4.

The second task turned out to be more interesting and more difficult. Polya increased the slope of the diagonal in Fig. 3.4 and found that in this case, the "diagonal sums" generate a new numerical sequence: $1, 1, 1, 2, 3, 4, 9, 13, \ldots$ He proposed to study this numerical result and prove that this sequence can be calculated by the recurrent formula $G_n = G_{n-1} + G_{n-3}$ and can also be expressed through binomial coefficients. Then he suggested further increasing the slope of the diagonal and getting the general formula for the "diagonal sums".

### 3.3.2.  Pascal's rightangular triangle

As it is known, there are many different forms of representation of Pascal's triangle. To solve the tasks, formulated by George Polya [111], in Stakhov's book [16] another table of representing the

Table 3.1. Pascal's rightangular triangle (numerical form).

| $m \downarrow, n \rightarrow$ | 0 | 1 | 2 | 3 | 4 | 5 | 6 | 7 | 8 | 9 |
|---|---|---|---|---|---|---|---|---|---|---|
| 0 | 1 | 1 | 1 | 1 | 1 | 1 | 1 | 1 | 1 | 1 |
| 1 | | 1 | 2 | 3 | 4 | 5 | 6 | 7 | 8 | 9 |
| 2 | | | 1 | 3 | 6 | 10 | 15 | 21 | 28 | 36 |
| 3 | | | | 1 | 4 | 10 | 20 | 35 | 56 | 84 |
| 4 | | | | | 1 | 5 | 15 | 35 | 70 | 126 |
| 5 | | | | | | 1 | 6 | 21 | 56 | 126 |
| 6 | | | | | | | 1 | 7 | 28 | 84 |
| 7 | | | | | | | | 1 | 8 | 36 |
| 8 | | | | | | | | | 1 | 9 |
| 9 | | | | | | | | | | 1 |
| | 1 | 2 | 4 | 8 | 16 | 32 | 64 | 128 | 512 | 1024 |

binomial coefficients in form, resembling a rightangular triangle (see Table 3.1), was suggested. Such a table of binomial coefficients was named *Pascal's rightangular triangle*.

As it follows from the example of *Pascal's rightangular triangle* (see Table 3.1), it has 10 columns $n = 0, 1, 2, 3, \ldots, 9$, numbered from left to right, and 10 rows $m = 0, 1, 2, 3, \ldots, 9$, numbered from top to down. The binomial coefficients are located in the form of a rightangular triangle, having the "horizontal leg" $\underbrace{1\,1\,1\ldots 1}_{10\,\text{units}}$, the "vertical leg" $\underbrace{1, 9, 36, 84, 126, 126, 84, 36, 9, 1}_{10\,\text{numbers}}$ and the "hypotenuse" $\underbrace{1\,1\,1\ldots 1}_{10\,\text{units}}$.

In future, we will also need an image of *Pascal's rightangular triangle* with the binomial coefficients, represented in the symbolic form (Table 3.2).

At the intersection of the $n$th column and the $m$th row of Table 3.2, the binomial coefficient $C_n^m$ stands. The "horizontal cathetus" in Table 3.2 is the *"zero"* row, which consists of the binomial coefficients $C_0^0 = C_1^0 = C_2^0 = \cdots = C_n^0 = 1$. The "vertical cathetus" in Table 3.2 is the $(n-1)$th, that is, the rightmost column, which consists of the following binomial coefficients (from top to bottom): $C_n^0, C_n^1, C_n^2, \ldots, C_n^n$. Let's note that the "hypotenuse" of Pascal's rightangular triangle (Table 3.2) consists of the following binomial coefficients: $C_0^0 = C_1^1 = C_2^2 = \cdots = C_n^n = 1$.

Table 3.2. Pascal's rightangular triangle (symbolic form).

| $m \downarrow, n \rightarrow$ | 0 | 1 | 2 | 3 | 4 | 5 | 6 | 7 | 8 | 9 |
|---|---|---|---|---|---|---|---|---|---|---|
| 0 | $C_0^0$ | $C_1^0$ | $C_2^0$ | $C_3^0$ | $C_4^0$ | $C_5^0$ | $C_6^0$ | $C_7^0$ | $C_8^0$ | $C_9^0$ |
| 1 | | $C_1^1$ | $C_2^1$ | $C_3^1$ | $C_4^1$ | $C_5^1$ | $C_6^1$ | $C_7^1$ | $C_8^1$ | $C_9^1$ |
| 2 | | | $C_2^2$ | $C_3^2$ | $C_4^2$ | $C_5^2$ | $C_6^2$ | $C_7^2$ | $C_8^2$ | $C_9^2$ |
| 3 | | | | $C_3^3$ | $C_4^3$ | $C_5^3$ | $C_6^3$ | $C_7^3$ | $C_8^3$ | $C_9^3$ |
| 4 | | | | | $C_4^4$ | $C_5^4$ | $C_6^4$ | $C_7^4$ | $C_8^4$ | $C_9^4$ |
| 5 | | | | | | $C_5^5$ | $C_6^5$ | $C_7^5$ | $C_8^5$ | $C_9^5$ |
| 6 | | | | | | | $C_6^6$ | $C_7^6$ | $C_8^6$ | $C_9^6$ |
| 7 | | | | | | | | $C_7^7$ | $C_8^7$ | $C_9^7$ |
| 8 | | | | | | | | | $C_8^8$ | $C_9^8$ |
| 9 | | | | | | | | | | $C_9^9$ |
| | 1 | 2 | 4 | 8 | 16 | 32 | 64 | 128 | 512 | 1024 |

It is clear that all the "internal" binomial coefficients in Tables 3.1 and 3.2 are calculated according to *Pascal's rule* (3.11), which for *Pascal's rightangular triangle* is formulated as follows: *an "internal" binomial coefficient $C_n^m$ (on the intersection of the nth column and the mth row) equal to the sum of the two binomial coefficients, the binomial coefficient $C_{n-1}^m$, which stands on the left of the binomial coefficient $C_n^m$ (on the intersection of the $(n-1)$th column and the mth row) and the binomial coefficient $C_{n-1}^{m-1}$, which stands over the binomial coefficient $C_n^m$ (on the intersection of the $(n-1)$th column and $(m-1)$th row), that is, $C_n^m = C_{n-1}^m + C_{n-1}^{m-1}$.*

Note also that in the $k$th column from top to bottom, the following binomial coefficients are located: $C_k^0, C_k^1, C_k^2, \ldots, C_k^n$; however, all cells under the "hypotenuse" are "empty". This means that all binomial coefficients of the type $C_n^m (m > n)$ are identically equal to "zero", that is,

$$C_n^m = 0 \quad \text{for } m > n. \tag{3.12}$$

If we sum up the binomial coefficients of the $n$th column of *Pascal's rectangular triangle*, then according to (3.8), we get the binary number $2^n$. If this is done for all columns, starting with zero row, then we get the well-known "binary sequence":

$$1, 2, 4, 8, 16, 32, 64, \ldots, 2^n, \ldots \tag{3.13}$$

These binary numbers are located in the bottom rows of Tables 3.1 and 3.2.

Thus, we can say that *Pascal's rectangular triangles* (see Tables 3.1 and 3.2) "generates" binary numbers!

### 3.3.3. Pascal's 1-triangle

Now let's shift each row of *Pascal's rectangular triangle* (Tables 3.1 and 3.2) on one column to the right relative to the previous row. As a result of this transformation, we get certain "deformed" Pascal's rectangular triangle called *Pascal's 1-triangle* (see Tables 3.3 and 3.4).

If we now sum up the binomial coefficients of Pascal's 1-triangles in Tables 3.3 and 3.4 by columns, then, to our amazement, we find that such summation leads us to the classical Fibonacci numbers (see the bottom row of Tables 3.3 and 3.4):

$$1, 1, 2, 3, 5, 8, 13, \ldots, F_1(n+1) = F_{n+1}, \ldots, \tag{3.14}$$

Table 3.3. Pascal's 1-triangle (numerical form).

|   | 0 | 1 | 2 | 3 | 4 | 5 | 6 | 7 | 8 | 9 |
|---|---|---|---|---|---|---|---|---|---|---|
| 0 | 1 | 1 | 1 | 1 | 1 | 1 | 1 | 1 | 1 | 1 |
| 1 |   |   | 1 | 2 | 3 | 4 | 5 | 6 | 7 | 8 |
| 2 |   |   |   |   | 1 | 3 | 6 | 10 | 15 | 21 |
| 3 |   |   |   |   |   |   | 1 | 4 | 10 | 20 |
| 4 |   |   |   |   |   |   |   |   | 1 | 5 |
|   | 1 | 1 | 2 | 3 | 5 | 8 | 13 | 21 | 34 | 55 |

Table 3.4. Pascal's 1-triangle (symbolic form).

|   | 0 | 1 | 2 | 3 | 4 | 5 | 6 | 7 | 8 | 9 |
|---|---|---|---|---|---|---|---|---|---|---|
| 0 | $C_0^0$ | $C_1^0$ | $C_2^0$ | $C_3^0$ | $C_4^0$ | $C_5^0$ | $C_6^0$ | $C_7^0$ | $C_8^0$ | $C_9^0$ |
| 1 |   |   | $C_1^1$ | $C_2^1$ | $C_3^1$ | $C_4^1$ | $C_5^1$ | $C_6^1$ | $C_7^1$ | $C_8^1$ |
| 2 |   |   |   |   | $C_2^2$ | $C_3^2$ | $C_4^2$ | $C_5^2$ | $C_6^2$ | $C_7^2$ |
| 3 |   |   |   |   |   |   | $C_3^3$ | $C_4^3$ | $C_5^3$ | $C_6^3$ |
| 4 |   |   |   |   |   |   |   |   | $C_4^4$ | $C_5^4$ |
|   | 1 | 1 | 2 | 3 | 5 | 8 | 13 | 21 | 34 | 55 |

where, through $F_1(n+1) = F_{n+1}$, we designated the $(n+1)$th Fibonacci number, which is located in the bottom row of the $n$th column of Pascal's 1-triangle (Tables 3.3 and 3.4); this Fibonacci number is determined by using the following recurrent formula:

$$F_1(n+1) = F_1(n) + F_1(n-1) \tag{3.15}$$

for the following initial conditions:

$$n > 2; \quad F_1(1) = F_2(1) = 1. \tag{3.16}$$

Now, we turn again to the book of Polya [111]. As mentioned above, in this book, Polya first interpreted the Fibonacci numbers through "diagonal sums" of Pascal's triangle and at the same time, he formulated the task *to express Fibonacci numbers $F_n$ through binomial coefficients*.

Note that *Pascal's rightangular triangle* (Tables 3.1 and 3.2) was used for the first time in Stakhov's book [16]. Such an approach proved to be very fruitful and led to the solution of the tasks formulated by Polya in [111].

In particular, the representation of Pascal's 1-triangle in symbolic form (Table 3.2) allowed solving very simply all George Polya's tasks formulated in [111].

Let's begin with the examples of the representation of the Fibonacci numbers as the sum of the binomial coefficients, which follow directly from Table 3.2:

$$
\begin{array}{|c|}
\hline
F_1 = F_1(1) = C_0^0 \\
\hline
F_2 = F_1(2) = C_1^0 \\
\hline
F_3 = F_1(3) = C_2^0 + C_1^1 \\
\hline
F_4 = F_1(4) = C_3^0 + C_2^1 \\
\hline
F_5 = F_1(5) = C_4^0 + C_3^1 + C_2^2 \\
\hline
F_6 = F_1(6) = C_5^0 + C_4^1 + C_3^2 \\
\hline
F_7 = F_1(7) = C_6^0 + C_5^1 + C_4^2 + C_3^3 \\
\hline
F_8 = F_1(8) = C_7^0 + C_6^1 + C_5^2 + C_4^3 \\
\hline
F_9 = F_1(9) = C_8^0 + C_7^1 + C_6^2 + C_5^3 + C_4^4 \\
\hline
F_{10} = F_1(10) = C_9^0 + C_8^1 + C_7^2 + C_6^3 + C_5^4 \\
\hline
\end{array}
\tag{3.17}
$$

The examples (3.17) lead to the following features of the representation of the Fibonacci numbers $F_1(n+1) = F_{n+1}$. If $n = 2k + 1$ ($k = 0, 1, 2, 3, \ldots$), then the Fibonacci numbers $F_1(2k+1) = F_{2k+1}$ can be represented as follows:

$$F_1(2k+1) = F_{2k+1} = C_{2k}^0 + C_{2k-1}^1 + C_{2k-2}^2$$
$$+ \cdots + C_{2k-(k-1)}^{k-1} + C_k^k. \tag{3.18}$$

If $n = 2k$ ($k = 0, 1, 2, 3, \ldots$), then the Fibonacci numbers $F_1(2k+2) = F_{2k+2}$ can be represented as follows:

$$F_1(2k+2) = F_{2k+2} = C_{2k+1}^0 + C_{2k}^1 + C_{2k-1}^2 + \cdots + C_{k+2}^{k-1} + C_{k+1}^k. \tag{3.19}$$

The formulas (3.18) and (3.19) can be combined as one mathematical formula:

$$F_1(n+1) = F_{n+1} = C_n^0 + C_{n-1}^1 + C_{n-2}^2 + \cdots + C_{k+r}^k, \tag{3.20}$$

where $n = 2k + r$, $k$ is quotient and $r$ is remainder of division of $n$ by 2.

This means that for the *even* $n = 2k$, the formula (3.20) reduces to the formula (3.18), and for the *odd* $n = 2k + 1$, it reduces to formula (3.19). The formulas (3.18)–(3.19) are the solution of the first task formulated by Polya in the book [111].

### 3.3.4. Pascal's *p*-triangles and Fibonacci *p*-numbers

The representation of Pascal's triangle in a symbolic rectangular form (Table 3.4) allows one to solve more complex tasks formulated by Polya in [111], namely, this representation leads us to an infinite set of new numerical sequences called the *Fibonacci p-numbers*, and also allows us to express them through binomial coefficients as it is proved in [16].

Let's consider the situation when in the initial Pascal's rectangular triangle (Table 3.1), we shift the binomial coefficients of each row on $p$ columns to the right relative to the previous row, where $p$ can take the values from the set $p \in \{0, 1, 2, 3, \ldots\}$. The

"deformed" Pascal's triangle, obtained in this way, will be called
*Pascal's p-triangle.*

It is clear that *Pascal's 0-triangle*, that is, Pascal's $p$-triangle,
corresponding to the case $p = 0$, is nothing but the initial *Pascal's
rectangular triangle* (Table 3.1). Pascal $p$-triangle, corresponding to
the case $p = 1$, is represented by Tables 2 and 3.

Let's consider *Pascal's p-triangles* corresponding to the case
$p = 2$ (Tables 3.5 and 3.6).

Now, let's summarize by columns the binomial coefficients of
Pascal's 2-triangles (Tables 3.5 and 3.6); as a result, we obtain a
new recurrent numerical sequence presented in the lower rows of
Tables 3.5 and 3.6:

$$1, \ 1, \ 1, \ 2, \ 3, \ 4, \ 6, \ 9, \ 13, \ 19, \ 28, \ 41, \ 60, \dots . \tag{3.21}$$

Let's denote the $n$th term of the sequence (3.21) by $F_2(n)$. It
is easy to see the following recurrent regularity in the numerical
sequence (3.21) that can be calculated by the recurrent formula:

$$F_2(n) = F_2(n-1) + F_2(n-3) \tag{3.22}$$

for the following initial conditions:

$$n > 3, \quad F_2(1) = F_2(2) = F_2(3) = 1. \tag{3.23}$$

Table 3.5. Pascal's 2-triangle (numerical form).

|   | 0 | 1 | 2 | 3 | 4 | 5 | 6 | 7 | 8 | 9 |
|---|---|---|---|---|---|---|---|---|---|---|
| 0 | 1 | 1 | 1 | 1 | 1 | 1 | 1 | 1 | 1 | 1 |
| 1 |   |   |   | 1 | 2 | 3 | 4 | 5 | 6 | 7 |
| 2 |   |   |   |   |   |   | 1 | 3 | 6 | 10 |
| 3 |   |   |   |   |   |   |   |   |   | 1 |
|   | 1 | 1 | 1 | 2 | 3 | 4 | 6 | 9 | 13 | 19 |

Table 3.6. Pascal's 2-triangle (symbolic form).

|   | 0 | 1 | 2 | 3 | 4 | 5 | 6 | 7 | 8 | 9 |
|---|---|---|---|---|---|---|---|---|---|---|
| 0 | $C_0^0$ | $C_1^0$ | $C_2^0$ | $C_3^0$ | $C_4^0$ | $C_5^0$ | $C_6^0$ | $C_7^0$ | $C_8^0$ | $C_9^0$ |
| 1 |   |   |   | $C_1^1$ | $C_2^1$ | $C_3^1$ | $C_4^1$ | $C_5^1$ | $C_6^1$ | $C_7^1$ |
| 2 |   |   |   |   |   |   | $C_2^2$ | $C_3^2$ | $C_4^2$ | $C_5^2$ |
| 3 |   |   |   |   |   |   |   |   |   | $C_3^3$ |
|   | 1 | 1 | 1 | 2 | 3 | 4 | 6 | 9 | 13 | 19 |

The sequence (3.21), calculated by the recurrent formula (3.22) for the initial conditions (3.23) will be called the *Fibonacci 2-numbers* [16].

In the general case (arbitrary $p$), by summing the binomial coefficients of Pascal's $p$-triangle by the columns, we get a recurrent numerical sequence, which for the given $p \in \{0, 1, 2, 3, \ldots\}$ is calculated by the following general recurrent relation:

$$F_p(n) = F_p(n-1) + F_p(n-p-1) \tag{3.24}$$

for the following initial conditions:

$$n > p+1, \quad F_p(1) = F_p(2) = \cdots = F_p(p+1) = 1. \tag{3.25}$$

The numerical sequences, defined by the formulas (3.24) and (3.25), will be called the *Fibonacci p-numbers* [16].

It is clear that, for the case $p = 0$, the recurrent formula (3.24) and the initial conditions (3.25) take the following forms:

$$F_0(n) = F_0(n-1) + F_0(n-1) = 2F_0(n-1), \tag{3.26}$$

$$F_0(1) = 1. \tag{3.27}$$

It is also clear that the recurrent formula (3.26) for the initial condition (3.27) "generates" binary numbers 1, 2, 4, 8, 16, 32, 64, 128, ..., which represent the extreme particular case of the Fibonacci $p$-numbers corresponding to $p = 0$.

Now, let's consider the case of $p = 1$. For this case, the formulas (3.24) and (3.25) are reduced to the formulas (3.15) and (3.16), which define the classical Fibonacci numbers (3.14).

Finally, let's find the Fibonacci $p$-sequence for the case of $p = \infty$. It is clear that, for this case, the Fibonacci $p$-numbers are given only by the formula (3.25), that is, the Fibonacci $p$-numbers, corresponding to $p = \infty$, are the infinite numerical sequence, which consists only of 1's.

By using the representation of the Pascal $p$-triangle in symbolic form (similar to Tables 3.4 and 3.6), it is easy to derive the formula, which allows representing the Fibonacci $p$-numbers through binomial coefficients in the explicit form (see [16]):

$$F_p(n+1) = C_n^0 + C_{n-p}^1 + C_{n-2p}^2 + C_{n-3p}^3 + \cdots + C_{k+r}^k, \qquad (3.28)$$

where $n = k(p+1) + r$, $k$ is quotient and $r$ is remainder of the division $n$ on $p+1$.

We note that the formula (3.28) is a solution of the general task formulated by Polya in [111] for the diagonal sums of *Pascal's triangle*.

### 3.4. The Extended Fibonacci *p*-Numbers

Until now, we have considered the sequences of the Fibonacci $p$-numbers $F_p(n)$ corresponding to the positive values of $n$. Let's consider the Fibonacci $p$-numbers $F_p(n)$ for the cases of $n = 0, -1, -2, -3, \ldots$ and extend them towards the negative values of $n$ in order to establish some common properties of such *extended* sequences. To calculate the Fibonacci $p$-numbers $F_p(0)$, $F_p(-1), \ldots, F_p(-p)$, corresponding to the values of $n = 0$, $-1, -2, -3, \ldots$, we will use the recurrent relation (3.24) and the initial conditions (3.25).

By representing the Fibonacci $p$-number $F_p(p+1)$ in the form (3.24), we get

$$F_p(p+1) = F_p(p) + F_p(0). \qquad (3.29)$$

Since, according to (3.25), $F_p(p) = F_p(p+1) = 1$, it follows from (3.29) that

$$F_p(0) = 0, \qquad (3.30)$$

and this statement is true for arbitrary integer $p > 0$.

By continuing this process, that is, by representing the Fibonacci $p$-numbers $F_p(p)$, $F_p(p-1)$, $F_p(p-2), \ldots, F_p(2)$ in the form (3.24), it is easy to prove that

$$F_p(0) = F_p(-1) = F_p(-2), \ldots, F_p(-p+1) = 0. \qquad (3.31)$$

Let's now represent the number $F_p(1)$ in the form:

$$F_p(1) = F_p(0) + F_p(-p). \tag{3.32}$$

Since according to (3.25), $F_p(1) = 1$ and according to (3.30), $F_p(0) = 0$, it follows from (3.32) that

$$F_p(-p) = 1. \tag{3.33}$$

Consistently by representing the Fibonacci $p$-numbers $F_p(0)$, $F_p(-1)$, $F_p(-2), \ldots$ in the form (3.24), it is easy to prove that

$$F_p(-p-1) = F_p(-p-2) = \cdots = F_p(-2p+1) = 0. \tag{3.34}$$

By continuing this process, we can get all the values of the Fibonacci $p$-numbers $F_p(n)$ for the negative values of $n$. Table 3.7 gives the idea of the extended Fibonacci $p$-numbers for the values $p = 1, 2, 3, 4, 5$.

Thus, the "manipulations" with Pascal's triangle lead us to a small mathematical discovery. We have discovered an infinite number of new numerical sequences, called the Fibonacci $p$-numbers ($p \in \{0, 1, 2, 3, \ldots\}$), given by the recurrent formula (3.24) for the initial conditions (3.25). These numerical sequences include the *binary sequence* (the case of $p = 0$) and the *classical Fibonacci numbers* (the case $p = 1$). These numerical sequences have a number of interesting mathematical properties, and their study leads to the expansion of the classical theory of the Fibonacci numbers [7–9, 11].

Table 3.7. The extended Fibonacci $p$-numbers.

| $n$ | 8 | 7 | 6 | 5 | 4 | 3 | 2 | 1 | 0 | −1 | −2 | −3 | −4 | −5 | −6 | −7 | −8 |
|---|---|---|---|---|---|---|---|---|---|---|---|---|---|---|---|---|---|
| $F_1(n)$ | 21 | 13 | 8 | 5 | 3 | 2 | 1 | 1 | 0 | 1 | −1 | 2 | −3 | 5 | −8 | 13 | −21 |
| $F_2(n)$ | 9 | 6 | 4 | 3 | 2 | 1 | 1 | 1 | 0 | 0 | 1 | 0 | −1 | 1 | 1 | −2 | 0 |
| $F_3(n)$ | 5 | 4 | 3 | 2 | 1 | 1 | 1 | 1 | 0 | 0 | 0 | 1 | 0 | 0 | −1 | 1 | 0 |
| $F_4(n)$ | 4 | 3 | 2 | 1 | 1 | 1 | 1 | 1 | 0 | 0 | 0 | 0 | 1 | 0 | 0 | 0 | −1 |
| $F_5(n)$ | 3 | 2 | 1 | 1 | 1 | 1 | 1 | 1 | 0 | 0 | 0 | 0 | 0 | 1 | 0 | 0 | 0 |

### 3.5. Generalization of the Golden Section Problem

### 3.5.1. The limit of the ratio of the neighboring Fibonacci $p$-numbers

It is known that the ratio of the neighboring classical Fibonacci numbers in the limit tends to the golden ratio, that is,

$$\lim_{n \to \infty} \frac{F_n}{F_{n-1}} = \Phi = \frac{1 + \sqrt{5}}{2}. \tag{3.35}$$

It is believed that the formula (3.35) was proven by the great astronomer and mathematician Johannes Kepler and therefore, in his honor, it is called the *Kepler's formula*.

Let's now consider the limit of the ratio of the neighboring Fibonacci $p$-numbers $\frac{F_p(n)}{F_p(n-1)}$ for $n \to \infty$. For this purpose, we denote this limit by $x$, i.e.,

$$\lim_{n \to \infty} \frac{F_p(n)}{F_p(n-1)} = x. \tag{3.36}$$

Let's now represent the ratio of the neighboring Fibonacci $p$-numbers as follows:

$$\frac{F_p(n)}{F_p(n-1)} = \frac{F_p(n-1) + F_p(n-p-1)}{F_p(n-1)}$$

$$= 1 + \frac{1}{\frac{F_p(n-1)}{F_p(n-p-1)}} = 1 + \frac{1}{\frac{F_p(n-1)F_p(n-2)\cdots F_p(n-p)}{F_p(n-2)F_p(n-3)\cdots F_p(n-p-1)}}. \tag{3.37}$$

Taking into consideration the definition (3.36), for the case $n \to \infty$, the equality (3.37) can be represented as follows:

$$x = 1 + \frac{1}{x^p},$$

whence the next algebraic equation follows:

$$x^{p+1} - x^p - 1 = 0. \tag{3.38}$$

Note that equation (3.38) defines the infinite number of algebraic equations because each $p$ corresponds to its own algebraic equation of the type (3.38). Denote by $\Phi_p$ the positive root of the algebraic equation (3.38).

In particular, for the case of $p = 0$, equation (3.38) reduces to the trivial equation $x = 2$, whose root is $\Phi_0 = 2$.

For the case of $p = 1$, equation (3.38) reduces to the *golden section equation*:

$$x^2 - x - 1 = 0, \tag{3.39}$$

whose root is the golden proportion $\Phi = \frac{1+\sqrt{5}}{2}$.

### 3.5.2. Geometric definition

Thus, the algebraic equation (3.38) is a generalization of the golden section equation (3.39). We note that equation (3.38) can also be obtained as a result of solving the geometric task in Fig. 3.5.

Let the integer $p$ take the following values $p = 0, 1, 2, 3, \ldots$ and the segment $AB$ be divided at the point $C$ in the following proportion (Fig. 3.5):

$$\frac{CB}{AC} = \left(\frac{AB}{CB}\right)^p. \tag{3.40}$$

Let's denote the value of the desired relation $\frac{AB}{CB}$ by $x$, that is,

$$\frac{AB}{CB} = x. \tag{3.41}$$

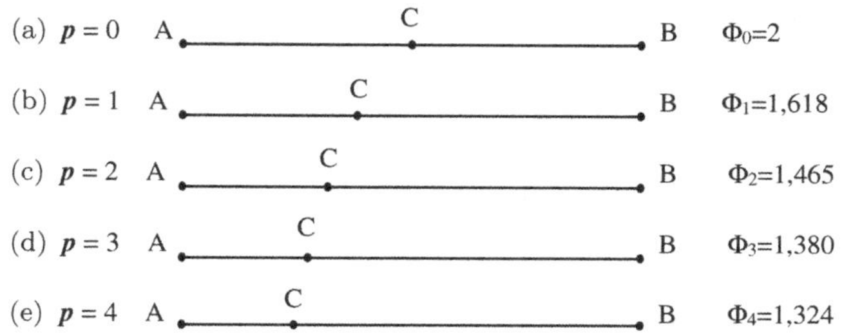

(a) $p = 0$   A —————— C —————— B   $\Phi_0=2$

(b) $p = 1$   A —————— C ———— B   $\Phi_1=1,618$

(c) $p = 2$   A ———— C —————— B   $\Phi_2=1,465$

(d) $p = 3$   A ——— C —————— B   $\Phi_3=1,380$

(e) $p = 4$   A —— C —————— B   $\Phi_4=1,324$

Fig. 3.5. The golden $p$-sections ($p = 0, 1, 2, 3, \ldots$).

Taking into consideration that $AB = AC + CB$, we can write

$$\frac{AB}{CB} = \frac{AC + CB}{CB} = 1 + \frac{1}{\frac{CB}{AC}}. \tag{3.42}$$

Taking into consideration the expressions (3.40) and (3.41), the expression (3.42) can be represented as follows:

$$x = 1 + \frac{1}{x^p},$$

whence we get the algebraic equation (3.38) obtained from the study of the limit of the ratio of the neighboring Fibonacci $p$-numbers given by (3.38).

Note that the division of the segment $AB$ at the point $C$, given by the proportion (3.40), reduces to the *dichotomy* for the case of $p = 0$ (Fig. 3.5(a)) and to the *classical golden section* for the case $p = 1$ (Fig. 3.5(b)). Taking into consideration this fact, the division of the segment $AB$ at the point $C$, given by the proportion (3.40), was called in [16] as the *golden p-section*, and the positive root of equation (3.38) by the *golden p-proportion* $\Phi_p$.

### 3.5.3. Some algebraic properties of the golden p-proportion

By representing equation (3.38) in the form $x^{p+1} = x^p + 1$ and by substituting $x$ with the golden $p$-proportion $\Phi_p$ in this equation, we get the following identity for the *golden p-proportion*:

$$\Phi_p^{p+1} = \Phi_p^p + 1. \tag{3.43}$$

By dividing all terms of the identity (3.43) by $\Phi_p^p$, we get the following identities for the *golden p-proportion* $\Phi_p$:

$$\Phi_p = 1 + \frac{1}{\Phi_p^p} \tag{3.44}$$

or

$$\Phi_p - 1 = \frac{1}{\Phi_p^p}. \tag{3.45}$$

Note that for the case $p = 0$, we have $\Phi_0 = 2$; then the identities (3.44) and (3.45) are reduced to the following trivial formulas:

$$2 = 1 + \frac{1}{1} \quad \text{or} \quad 2 - 1 = \frac{1}{1}.$$

For the case $p = 1$, we have $\Phi_1 = \Phi = \frac{1+\sqrt{5}}{2}$; then the identities (3.44) and (3.45) are reduced to the following well-known identities for the classical golden proportion:

$$\Phi = 1 + \frac{1}{\Phi} \quad \text{or} \quad \Phi - 1 = \frac{1}{\Phi}.$$

We will now multiply and divide many times all the terms of the identity (3.43) by the golden $p$-proportion $\Phi_p$; as a result, we obtain the following general identities, connecting the degrees of the *golden p-proportions*:

$$\Phi_p^n = \Phi_p^{n-1} + \Phi_p^{n-p-1} = \Phi_p \times \Phi_p^{n-1}, \quad n = 0, \pm 1, \pm 2, \pm 3, \ldots .$$
$$(3.46)$$

Note that for the case $p = 0 (\Phi_0 = 2)$, the identities (3.46) are reduced to the following trivial identities for the binary numbers:

$$2^n = 2^{n-1} + 2^{n-1} = 2 \times 2^{n-1}, \quad n = 0, \pm 1, \pm 2, \pm 3, \ldots . \quad (3.47)$$

For the case of $p = 1$, we have $\Phi_1 = \Phi = \frac{1+\sqrt{5}}{2}$; for this case, the identities (3.46) are reduced to the following trivial identities for the classical golden proportion:

$$\Phi^n = \Phi^{n-1} + \Phi^{n-2} = \Phi \times \Phi^{n-1}, \quad n = 0, \pm 1, \pm 2, \pm 3, \ldots . \quad (3.48)$$

## 3.6. Algebraic Equations for the Golden $p$-Proportion and Vieta's Formulas

### 3.6.1. General properties of the roots of the golden $p$-proportion equations

Let's turn once again to the equation of the golden $p$-proportions (3.38). In [80], the following properties of the roots of this algebraic equation was studied and the following general properties of these roots were deduced:

(1) In accordance with Descartes' *rule of the sign*, all equations of the golden $p$-proportions of the kind (3.38) have only one positive root, the *golden p-proportion* $\Phi_p$.

(2) The equation of the golden $p$-proportion (3.38) has $(p + 1)$th degrees. This means that equation (3.38) has $p + 1$ roots $x_1, x_2, \ldots, x_{p+1}$. In future, without loss of generality, we assume that the root $x_1$ always coincides with the *golden p-proportion*, that is, $x_1 = \Phi_p$.

(3) Because all coefficients 1, $-1$ and $-1$ of equation (3.38) are real numbers, this means that complex roots arise in pairs, that is, each complex root $a + bi$ always occurs together with the complex conjugate root $a - bi$.

(4) We can represent the polynomial of the golden $p$-proportion as a product of binomials:

$$x^{p+1} - x^p - 1 = (x - x_1)(x - x_2) \cdots (x - x_p)(x - x_{p+1}).$$
$$(3.49)$$

(5) For every root $x_k (k = 1, 2, 3, \ldots, p + 1)$, we have the following identity:

$$x_k^n = x_k^{n-1} + x_k^{n-p-1} = x_k \times x_k^{n-1}, \qquad (3.50)$$

where $n = 0, \pm 1, \pm 2, \pm 3, \ldots$.

Note that the identity (3.46) for the *golden p-proportion* $\Phi_p$ is the partial case of the general identity (3.50).

### 3.6.2. Vieta's formulas for quadratic equations

The French mathematician Francois Vieta (1540–1603) made a great contribution to the development of the *elementary algebra*. He deduced the so-called *Vieta's formulas*, which give connection between the coefficients of a polynomial and its roots (Fig. 3.6).

Fig. 3.6. Francois Vieta (1540–1603).

Let's consider *Vieta's formulas* for the quadratic equation. If $x_1$, $x_2$ are the roots of the quadratic equation:

$$ax^2 + bx + c = 0, \tag{3.51}$$

then there are the following relations, connecting the roots of the equation with its coefficients $a$, $b$, $c$:

$$x_1 + x_2 = -\frac{b}{a}, \quad x_1 x_2 = +\frac{c}{a} \tag{3.52}$$

In particular, if $a = 1$, equation (3.51) takes the form of the *reduced equation*:

$$x^2 + px + q = 0, \tag{3.53}$$

for which the following relations hold:

$$x_1 + x_2 = -p, \quad x_1 x_2 = +q. \tag{3.54}$$

Thus, the *Vieta's rule* says: *The sum of the roots of the reduced quadratic equation* (3.53) *is equal to the second coefficient with the opposite sign, and the product of the roots is equal to the free term.*

It is convenient to use this rule to check the solution of the quadratic equation as well as to compose the quadratic equation for its given roots.

### 3.6.3. Vieta's formulas for the general case

However, *Vieta's formulas* for general case are the most interesting. Let's consider these formulas. If $x_1, x_2, \ldots, x_n$ are the roots of the *reduced algebraic equation*

$$x^n + a_1 x^{n-1} + a_2 x^{n-2} + \cdots + a_n = 0, \qquad (3.55)$$

then the coefficients $a_1, a_2, \ldots, a_n$ are expressed through the roots of equation (3.55) as follows:

$$a_1 = -(x_1 + x_2 + \cdots + x_n)$$

$$a_2 = x_1 x_2 + x_1 x_3 + \cdots + x_1 x_n + x_2 x_3 + \cdots + x_{n-1} x_n$$

$$a_3 = -(x_1 x_2 x_3 + x_1 x_2 x_4 + \cdots + x_{n-2} x_{n-1} x_n)$$

$$\vdots \qquad\qquad (3.56)$$

$$a_{n-1} = (-1)^{n-1}(x_1 x_2 \cdots x_{n-1} + x_1 x_2 \ldots x_{n-2} x_n$$

$$+ \cdots + x_2 x_3 \ldots x_n)$$

$$a_n = (-1)^n x_1 x_2 \ldots x_n$$

### 3.6.4. Vieta's formulas for the roots of the golden section equation

The golden section equation is the *reduced equation* of the type (3.53) with the coefficients:

$$p = -1, \quad q = -1. \qquad (3.57)$$

Then, according to the *Vieta's rule* of (3.54), the following relations exist between the roots of equation (3.53):

$$x_1 + x_2 = -p = -(-1) = 1, \quad x_1 x_2 = +q = -1. \qquad (3.58)$$

### 3.6.5. Vieta's formulas for the equation of the golden p-proportion

In general, the reduced algebraic equation of $(p+1)$th degree has the following form:

$$x^{p+1} + a_1 x^p + a_2 x^{p-1} + \cdots + a_p x + a_{p+1} = 0. \qquad (3.59)$$

By comparing equation (3.59) to the algebraic equation (3.38), we can write the following values for the coefficients of the algebraic equation (3.59) for the case of equation (3.38):

$$a_1 = -1, \quad a_2 = a_3 = \cdots = a_p = 0, \quad a_{p+1} = -1. \qquad (3.60)$$

By using *Vieta's formulas* (3.56) with regards to (3.60), we can write the following formulas that relate the coefficients (3.60) to the roots $x_1, x_2, \ldots, x_{p+1}$ of equation (3.38):

$$x_1 + x_2 + x_3 + \cdots + x_p + x_{p+1} = 1 \qquad (3.61)$$

$$(x_1 x_2 + x_1 x_3 + \cdots + x_1 x_{p+1}) + (x_2 x_3 + \cdots + x_2 x_{p+1})$$
$$+ \cdots + x_p x_{p+1} = 0$$

$$(x_1 x_2 x_3 + x_1 x_3 x_4 + \cdots + x_1 x_p x_{p+1}) + (x_2 x_3 x_4 + \cdots + x_2 x_p x_{p+1})$$
$$+ \cdots + x_{p-1} x_p x_{p+1} = 0$$

$$\vdots \qquad (3.62)$$

$$x_1 x_2 \ldots x_{p-2} x_{p-1} x_p + x_1 x_3 x_4 \ldots x_{p-1} x_p x_{p+1}$$
$$+ \cdots + x_2 x_3 x_4 \ldots x_{p-1} x_p x_{p+1} = 0,$$

$$x_1 x_2 x_3 \ldots x_{p-1} x_p x_{p+1} = (-1)^p. \qquad (3.63)$$

As it follows from (3.59) and (3.61), the sum of all the roots of equation (3.38) is identically equal to 1, and their product is equal to $(+1)$ (for the *even* values of $p$) or $(-1)$ (for the *odd* values of $p$).

The following theorem has been proved in [80].

**Theorem 3.1.** *For any integer $p = 1, 2, 3, \ldots$ for the case, when $k$ takes the values from the set $k \in \{1, 2, 3, \ldots, p\}$, the following identity* *is true*:

$$(x_1 + x_2 + x_3 + \cdots + x_{p+1})^k = x_1^k + x_2^k + x_3^k + \cdots + x_{p+1}^k = 1.$$
(3.64)

## 3.7. Binet's Formulas for the Fibonacci $p$-Numbers

### 3.7.1. General approach

In Chapter 2, Binet's formulas for the classic Fibonacci and Lucas numbers had been found. They allowed one to express the Fibonacci and Lucas numbers through the roots $x_1 = \frac{1+\sqrt{5}}{2}$ and $x_2 = \frac{1-\sqrt{5}}{2}$ of the golden section equation: $x^2 - x - 1 = 0$.

Note that the following expression was used as the source for the derivation of *Binet's formula* for the Fibonacci numbers:

$$F_n = k_1 x_1^n - k_2 x_2^n,$$
(3.65)

where the constant coefficients $k_1$ and $k_2$ are calculated by using the system of algebraic equations:

$$F_0 = k_1 - k_2 = 0,$$
$$F_1 = k_1 x_1 - k_2 x_2 = 1.$$
(3.66)

To derive *Binet's formula* for the Lucas numbers, we will use the following representation of the Lucas numbers through the roots $x_1 = \frac{1+\sqrt{5}}{2}$ and $x_2 = \frac{1-\sqrt{5}}{2}$:

$$L_n = x_1^n + x_2^n.$$
(3.67)

We use the same approach to derive Binet's formulas for the Fibonacci and Lucas $p$-numbers. We begin our reasoning from the Fibonacci $p$-numbers.

For any given integer $p = 1, 2, 3, \ldots$, the extended *Fibonacci p-number* $F_p(n)(n = 0, \pm 1, \pm 2, \pm 3, \ldots)$ can be represented as follows:

$$F_p(n) = k_1 x_1^n + k_2 x_2^n + \cdots + k_{p+1} x_{p+1}^n,$$
(3.68)

where $x_1, x_2, \ldots, x_{p+1}$ are the roots of the algebraic equation (3.38) $x^{p+1} - x^p - 1 = 0$, and $k_1, k_2, \ldots, k_{p+1}$ are some constants, which are calculated by using the following system of the algebraic equations:

$$F_p(0) = k_1 + k_2 + \cdots + k_{p+1} = 0$$
$$F_p(1) = k_1 x_1 + k_2 x_2 + \cdots + k_{p+1} x_{p+1} = 1$$
$$F_p(2) = k_1 x_1^2 + k_2 x_2^2 + \cdots + k_{p+1} x_{p+1}^2 = 1 \qquad (3.69)$$
$$\vdots$$
$$F_p(p) = k_1 x_1^p + k_2 x_2^p + \cdots + k_{p+1} x_{p+1}^p = 1.$$

The formula (3.68), in which the constant coefficients $k_1, k_2, \ldots, k_{p+1}$ are calculated by using the system of the algebraic equations (3.69), is *Binet's formula* for the *Fibonacci p-numbers*.

## 3.8. Binet's Formulas for the Lucas $p$-Numbers

Note that the following reasonings were used in Chapter 2 as the source for the derivation of *Binet's formula* for the *classic Lucas numbers*. We represented the extended classic Lucas numbers $L_n (n = 0, \pm 1, \pm 2, \pm 3, \ldots)$ in the form $L_n = x_1^n + x_2^n$. For the cases $n = 0$ and $n = 1$, we used the following initial values of the extended Lucas numbers: $L_0 = 2$ and $L_1 = 1$.

In general case, for any given integers $p = 1, 2, 3, \ldots$ and $n = 0, \pm 1, \pm 2, \pm 3, \ldots$, the formula

$$L_p(n) = x_1^n + x_2^n + \cdots + x_{p+1}^n \qquad (3.70)$$

sets forth a new class of recurrent numerical sequences $L_p(n)$, called *Lucas p-numbers*, which, in turn, are given by the following recurrent relation:

$$L_p(n) = L_p(n-1) + L_p(n-p-1) \qquad (3.71)$$

for the initial conditions:

$$L_p(0) = p+1; \quad L_p(1) = L_p(2) = \cdots = L_p(p) = 1. \qquad (3.72)$$

We begin with the proof of the initial conditions (3.72). Indeed, when $n = 0$, the expression (3.70) takes the following form:

$$L_p(0) = x_1^0 + x_2^0 + \cdots + x_{p+1}^0 = p + 1.$$

To calculate the remaining values of the initial *Lucas p-numbers* from (3.72), we use Theorem 3.1, according to which for any $k \in \{1, 2, 3, \ldots, p\}$, the following identity is true for the roots $x_1, x_2, x_3, \ldots, x_{p+1}$ of equation (3.38) $x^{p+1} - x^p - 1 = 0$:

$$x_1^k + x_2^k + x_3^k + \cdots + x_{p+1}^k = 1. \tag{3.73}$$

If we now represent the *Lucas p-numbers* $L_p(1)$, $L_p(2), \ldots, L_p(p)$ in the form (3.70), then we get the following representations through the roots $x_1, x_2, x_3, \ldots, x_{p+1}$:

$$L_p(1) = x_1 + x_2 + \cdots + x_{p+1}$$
$$L_p(2) = x_1^2 + x_2^2 + \cdots + x_{p+1}^2$$
$$\vdots \tag{3.74}$$
$$L_p(p) = x_1^p + x_2^p + \cdots + x_{p+1}^p.$$

By comparing (3.74) to (3.73), we can conclude

$$L_p(1) = L_p(2) = \cdots = L_p(p) = 1. \tag{3.75}$$

Now, we prove the validity of the recurrent formula (3.71) for the general case. To do this, it suffices to use the expression (3.70) and represent the expression (3.71) as follows:

$$L_p(n) = (x_1^{n-1} + x_2^{n-1} + \cdots + x_{p+1}^{n-1})$$
$$+ (x_1^{n-p-1} + x_2^{n-p-1} + \cdots + x_{p+1}^{n-p-1})$$
$$= L_p(n-1) + L_p(n-p-1), \tag{3.76}$$

whence the validity of the recurrent formula (3.71) for the *positive* values of $n$ follows.

For the *negative* values of $n$, the recurrent formula (3.71) is easily proved if we use the formula (3.70).

### 3.8.1. Binet's formula for the Fibonacci 2-numbers

Let's give the number $p = 2$ and apply the above approach to derive *Binet's formula* for the *Fibonacci 2-numbers*. For the case of $p = 2$, the recurrent relation (3.71) $F_p(n) = F_p(n-1) + F_p(n-p-1)$, the initial conditions (3.72) $F_p(1) = F_p(2) = \cdots = F_p(p+1) = 1$ and the algebraic equation (3.38) take the following form:

$$F_2(n) = F_2(n-1) + F_2(n-3), \tag{3.77}$$

$$F_p(1) = F_p(2) = F_p(3) = 1, \tag{3.78}$$

$$x^3 - x^2 - 1 = 0. \tag{3.79}$$

The algebraic equation (3.79) has three roots, a real root $x_1$ and two complex-conjugate roots $x_2$, $x_3$, which are given by the following expressions:

$$x_1 = \frac{h}{6} + \frac{2}{3h} + \frac{1}{3} = 1.4655712319\ldots,$$

$$x_2 = -\frac{h}{12} - \frac{1}{3h} + \frac{1}{3} - i\frac{\sqrt{3}}{2}\left(\frac{h}{6} - \frac{2}{3h}\right) = -0.233\ldots - (0.793\ldots)i,$$

$$x_3 = -\frac{h}{12} - \frac{1}{3h} + \frac{1}{3} + i\frac{\sqrt{3}}{2}\left(\frac{h}{6} - \frac{2}{3h}\right) = -0.233\ldots + (0.793\ldots)i,$$

$$\tag{3.80}$$

where

$$h = \sqrt[3]{116 + 12\sqrt{93}}.$$

By using the general expression (3.68), we can write *Binet's formula* for Fibonacci 2-numbers as follows:

$$F_2(n) = k_1 x_1^n + k_2 x_2^n + k_3 x_3^n. \tag{3.81}$$

The system of the algebraic equations (3.74) takes the following form:

$$F_2(0) = k_1 + k_2 + k_3 = 0,$$

$$F_2(1) = k_1 x_1 + k_2 x_2 + k_3 x_3 = 1, \tag{3.82}$$

$$F_3(2) = k_1 x_1^2 + k_2 x_2^2 + k_3 x_3^2 = 1.$$

By solving the system of equations (3.82), we get

$$k_1 = \frac{2h(h+2)}{h^3 + 8}, \tag{3.83}$$

$$k_2 = \frac{[-(h+2) + i\sqrt{3}(h-2)]h}{h^3 + 8}, \tag{3.84}$$

$$k_3 = \frac{[-(h+2) - i\sqrt{3}(h-2)]h}{h^3 + 8}, \tag{3.85}$$

where

$$h = \sqrt[3]{116 + 12\sqrt{93}}. \tag{3.86}$$

Note that the number (3.86) is an irrational number. Therefore, the coefficients $k_1$, $k_2$, $k_3$, specified by (3.83)–(3.85), are also irrational numbers. Moreover, the coefficients $k_2$, $k_3$, given by (3.84) and (3.85), are the complex conjugate numbers of the type $a + bi$ and $a - bi$, and the coefficients $a$ and $b$ for this case are irrational numbers.

By using the expressions (3.83)–(3.85), we can write *Binet's formula* for the Fibonacci 2-numbers as follows:

$$\begin{aligned}
F_2(n) = {} & \frac{2h(h+2)}{h^3 + 8} \times \left( \frac{h}{6} + \frac{2}{3h} + \frac{1}{3} \right)^n \\
& + \frac{(-(h+2) + i\sqrt{3}(h-2))h}{h^3 + 8} \\
& \times \left( -\frac{h}{12} - \frac{1}{3h} + \frac{1}{3} + i\frac{\sqrt{3}}{2} \left( \frac{h}{6} - \frac{2}{3h} \right) \right)^n \\
& + \frac{(-(h+2) - i\sqrt{3}(h-2))h}{h^3 + 8} \\
& \times \left( -\frac{h}{12} - \frac{1}{3h} + \frac{1}{3} - i\frac{\sqrt{3}}{2} \left( \frac{h}{6} - \frac{2}{3h} \right) \right)^n. \tag{3.87}
\end{aligned}$$

### 3.8.2. Binet's formula for the Lucas 2-numbers

By using the general formula (3.70), we can write *Binet's formula* for the Lucas 2-numbers as follows:

$$L_2(n) = x_1^n + x_2^n + x_3^n. \qquad (3.88)$$

By substituting into (3.88) the values of the roots $x_1$, $x_2$, $x_3$, given by the expressions (3.80), after simple transformations, we obtain *Binet's formula* for the Lucas 2-numbers in the form:

$$L_2(n) = \left(\frac{h}{6} + \frac{2}{3h} + \frac{1}{3}\right)^n + \left(-\frac{h}{12} - \frac{1}{3h} + \frac{1}{3} + i\frac{\sqrt{3}}{2}\left(\frac{h}{6} - \frac{2}{3h}\right)\right)^n$$

$$+ \left(-\frac{h}{12} - \frac{1}{3h} + \frac{1}{3} - i\frac{\sqrt{3}}{2}\left(\frac{h}{6} - \frac{2}{3h}\right)\right)^n. \qquad (3.89)$$

We now calculate the initial values of the sequence $L_2(n)$, given by (3.89). For the case $n = 0$, it directly follows from (3.89):

$$L_2(0) = 3. \qquad (3.90)$$

For cases $n = 1, 2$, if we take into consideration (3.76), we get, from (3.87),

$$L_2(1) = L_2(2) = 1. \qquad (3.91)$$

At first glance, it seems incredible that *Binet's formulas* (3.88) and (3.89), which are very complicated combinations of the complex conjugate numbers with irrational coefficients, really define the integer Fibonacci and Lucas 2-numbers for any integers $n = 0, \pm 1, \pm 2, \pm 3, \ldots$. And here, we can only admire the mathematical method, which allows us to express the most complicated mathematical truths in a compact form.

# Chapter 4

# Platonic Solids: From Plato's Cosmology to Fullerenes and Quasicrystals

## 4.1. The Golden Section in Platonic Solids

A person shows interest in regular polygons and polyhedra throughout his entire conscious activity: from a two-year-old child, playing with wooden cubes, to a mature mathematician. Some of the regular and semi-regular solids are found in Nature in the form of crystals, others in the form of viruses that can be considered by using an electron microscope.

What are a polygon and a polyhedron? To answer this question, we recall that geometry itself is sometimes defined as the science of space and spatial figures, two-dimensional and three-dimensional. A two-dimensional figure can be defined as a set of straight lines, bounding a part of a plane. Such a flat shape is called a *polygon.* It follows from this that a *polyhedron* can be defined as a set of polygons, bounding a part of three-dimensional space. Polygons, forming a polyhedron, are called its *faces.*

Scientists have long been interested in the *ideal* or *regular polygons*, that is, the polygons, which have equal sides and equal angles. The simplest regular polygon is a *equilateral triangle* because it has the smallest number of sides that can restrict a part of the plane. In the general case, together with the *equilateral triangle*, we have the following regular polygons: *square* (four sides), *pentagon* (five sides), *hexagon* (six sides), *octagon* (eight sides), *decagon* (ten sides), etc. Obviously, theoretically, there are no restrictions on the number of sides of the *regular polygons*, that is, the number of regular polygons is infinite.

What is a regular polyhedron? Such polyhedrons are called *regular* if all their faces are equal (or congruent) to each other and at the same time, they are *regular polygons*. How many regular polyhedra exist? At first glance, the answer to this question is very simple: there exist regular polyhedra as much as many regular polygons exist. However, it is not so. In Euclid's *Elements*, we find a strict proof that there are only *five convex regular polyhedra*, and their faces can be only three types of regular polygons: *triangles*, *squares*, and *pentagons*.

### 4.1.1. Platonic solids

Many books are dedicated to the theory and practice of polyhedra. The book of English mathematician, M. Venninger's *Models of Polyhedra* [112], is the most famous of these. This book begins with a description of the so-called *regular polyhedra*, i.e., polyhedra, formed by the simplest regular polygons of the same type. These polyhedra are usually called *Platonic solids* (Fig. 4.1), named after the ancient Greek philosopher Plato, who used regular polyhedra in his cosmology.

We will continue our consideration of the *Platonic solids* from regular polyhedra, whose faces are *equilateral triangles*. The *tetrahedron* is the simplest among regular polyhedra. In the *tetrahedron*, three equilateral triangles meet at one vertex; at the same time, its bases form a new equilateral triangle. The tetrahedron has the smallest number of faces among the *Platonic solids* and is a three-dimensional analog of a flat equilateral triangle that has the smallest number of sides among regular polygons.

The *octahedron* is the next *Platonic solid*, whose faces are equilateral triangles. In the *octahedron*, the four equilateral triangles meet in one vertex; as a result, we get a pyramid with a quadrilateral

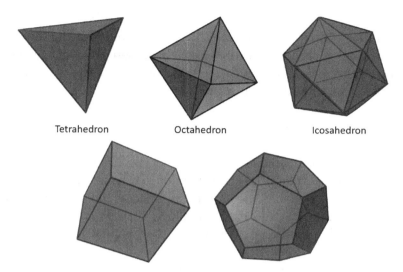

Tetrahedron          Octahedron          Icosahedron

Fig. 4.1. Platonic solids: tetrahedron, octahedron, cube, dodecahedron, and icosahedron.

base. If we connect two such pyramids by their bases, we get the *Platonic solid* with the eight triangular faces called an *octahedron*.

Now, we can try to connect at one point the five equilateral triangles. As a result, we obtain the *Platonic solid* with the 20 triangular faces called an *icosahedron*.

The *square* is the next regular form of polygon. If you connect three squares at one point and then connect such a form with three similar forms, we get the *Platonic solid* with six faces, which is called a *hexahedron* or a *cube*.

Finally, there is another possibility to construct a regular polyhedron based on the use of a *regular pentagon*. If we assemble 12 regular pentagons in such a way that at each point, three regular pentagons meet, then we obtain another *Platonic solid* called a *dodecahedron*.

The *hexagon* is the next regular polygon. However, if we connect three hexagons at one point, then we get a plane, i.e., it is impossible to construct a bulk figure from hexagons. Any other regular polygons with the number of sides more than 6 cannot form bulk figures at all.

In essence, we repeated the arguments of Euclid, conducted by him in Book XIII of the *Elements*. As it is well known, this book is dedicated to the presentation of the completed geometric theory of the *Platonic solids* (Fig. 4.1). It follows from these arguments that there are only five regular polyhedra, whose faces can only be *equilateral triangles*, *squares*, and *pentagons*.

There are amazing geometric relationships between regular polyhedra. *Duality property* is one of them. For example, the *cube* and the *octahedron* are dual to one another, i.e., they can be obtained from each other if we take the center of gravity of the faces of the first *Platonic solid* (the *cube*) as the vertices of the second *Platonic solid* (the *octahedron*) and back. Similarly, the *icosahedron* is dual to the *dodecahedron* and back. The *tetrahedron* is dual to itself.

### 4.1.2. Numerical parameters of the Platonic solids

The main numerical characteristics of the *Platonic solids* are as follows: the general *number of faces m*, the *number of faces n*, *convergent at each vertex*, the *total number of faces F*, the *number of vertices V*, the *number of edges E*, and the number of *flat angles A* on the surface of the polyhedron. Euler proved the famous formula:

$$V - E + F = 2,$$

connecting the number of vertices, edges and faces of arbitrary convex polyhedron.

The above numerical characteristics are given in Table 4.1.

### 4.1.3. The golden section in dodecahedron and icosahedron

The *dodecahedron* and the dual to it, *icosahedron*, occupy a special place among the *Platonic solids*. First of all, it is necessary to emphasize that the geometry of *dodecahedron* and *icosahedron* is directly connected with the *golden section*. Indeed, the faces of the *dodecahedron* are the *regular pentagons* based on the *golden section*. If we look closely at the *icosahedron*, we can see that at each of its vertices, there are five triangles, the outer sides of which form a

Table 4.1. Numerical Parameters of Platonic Solids.

| Polyherdron | Number of sides of the faces ($m$) | Number of faces, convergent in vertex ($n$) | Total number of faces ($F$) | Total number of vertices ($V$) | Total number of edges ($E$) | Total number of flat angles on surface ($A$) |
|---|---|---|---|---|---|---|
| Tetrahedron | 3 | 3 | 4 | 4 | 6 | 12 |
| Hexahedron (cube) | 4 | 3 | 6 | 8 | 12 | 24 |
| Octahedron | 3 | 4 | 8 | 6 | 12 | 24 |
| Icosahedron | 3 | 5 | 20 | 12 | 30 | 60 |
| Dodecahedron | 5 | 3 | 12 | 20 | 30 | 60 |

Table 4.2. The golden proportion in the spheres of dodecahedron and icosahedron.

| Polyhedron | $R_c$ | $R_m$ | $R_i$ |
|---|---|---|---|
| Icosahedron | $\frac{1}{2}\Phi\sqrt{3-\Phi}$ | $\frac{1}{2}\Phi$ | $\frac{\frac{1}{2}\Phi^2}{\sqrt{3}}$ |
| Dodecahedron | $\frac{\Phi\sqrt{3}}{2}$ | $\frac{\Phi^2}{2}$ | $\frac{\Phi^2}{2\sqrt{3-\Phi}}$ |

*regular pentagon.* Already these facts are enough to make sure that the *golden section* plays a decisive role in the construction of these two Platonic solids.

But there are deeper mathematical proofs of the fundamental role of the *golden section* in *icosahedron* and *dodecahedron*. It is known that these *Platonic solids* have three specific spheres. The first (inner) sphere is inscribed in the *Platonic solid* and touches its faces, and its radius is denoted by $R_i$. The second or middle sphere concerns its edges, and its radius is denoted by $R_m$. Finally, the third (outer) sphere is described around the *Platonic solids* and passes through its vertices, and its radius is denoted by $R_c$. It is proved in geometry that the values of the radii of these spheres for *icosahedron* and *dodecahedron*, having the edges of the unit length, can be expressed in terms of the *golden proportion* $\Phi$ (Table 4.2).

Note that the ratio of the radii $\frac{R_c}{R_i} = \frac{\sqrt{3(3-\Phi)}}{\Phi}$ are the same for icosahedron and dodecahedron. Thus, if *dodecahedron* and *icosahedron* have identical inscribed spheres, then their described spheres coincide. The proof of this mathematical result is given in Euclid's *Elements*. Thus, there are a huge number of relations, proved by the ancient mathematicians, which confirm the remarkable fact that the *golden proportion* is the main proportion of *dodecahedron* and *icosahedron*, and this fact is especially interesting from the point of view of the so-called *dodecahedron-icosahedral doctrine*, which we will consider below.

## 4.2. Archimedean Truncated Icosahedron and Stellate Polyhedra

### 4.2.1. Archimedean solids

There is a set of perfect solids, known as *Archimedean* or *semi-regular polyhedra*. All polyhedral angles of the Archimedean polyhedra are equal and all faces are regular polygons, but there are several different types. There are 13 semi-regular polyhedra attributed to Archimedes.

We will not analyze all the Archimedean solids and we refer our readers to Ref. 112. The greatest interest for us in the future will be to present the so-called *Archimedean truncated solids*, which are obtained from the *Platonic solids* by means of the truncation of their vertices. For the Platonic solids, their truncation can be done in such a way that the resulting new faces and the remaining parts of the *Platonic solids* are regular polygons. For example, a tetrahedron (Fig. 4.1) can be truncated so that its four triangular faces turn into four hexagonal ones, and four regular triangular faces are added to them. In this way, the following five Archimedean truncated solids can be obtained: *truncated tetrahedron*, *truncated hexahedron (cube)*, *truncated octahedron*, *truncated dodecahedron*, and *truncated icosahedron* (Fig. 4.2).

### 4.2.2. Archimedian truncated icosahedron

Among the five Archimedean truncated solids, the *truncated icosahedron* is of the greatest interest. This interest is associated with the discovery of *fullerenes*; for this discovery, a group of scientists (Robert Curl, Harold Kroto, Richard Smalley) were awarded the

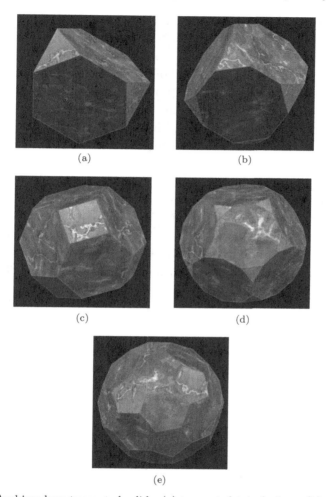

Fig. 4.2. Archimedean truncated solids: (a) truncated tetrahedron, (b) truncated cube, (c) truncated octahedron, (d) truncated dodecahedron, (e) truncated icosahedron.

Nobel Prize in Chemistry in 1996. In his Nobel lecture, the American scientist, Richard Smalley, one of the authors of the experimental discovery of fullerenes, speaks about Archimedes (287–212 BC) as the first researcher of the truncated polyhedra, in particular, the truncated icosahedron. However, he mentions that perhaps Archimedes appropriates this merit, and perhaps the icosahedrons were truncated long before him. It is enough to mention here about the discovery in Scotland, dated back to 2000 BC, of hundreds of stone objects (apparently for ritual purposes) in the form of spheres and various polyhedrons (solids, bounded on all sides by flat faces), including icosahedrons and dodecahedrons. Unfortunately, the original work of Archimedes has not been preserved, and its results have reached us, as they say, from "second-hand".

So, how to construct the Archimedean truncated icosahedron from the Platonic icosahedron? The answer is illustrated in Fig. 4.3. Indeed, as can be seen from Fig. 4.3, the five faces converge at any of the 12 vertices of the icosahedron. If we cut off each vertex of the Platonic icosahedron by the plane, then the 12 new pentagonal faces are formed. Together, with the 20 existing faces that had transformed from triangular to hexagonal ones after such cutting, the 32 faces of the truncated icosahedron will be formed. As a result, we get the Archimedean truncated icosahedron with 90 edges and 60 vertices (Fig. 4.4).

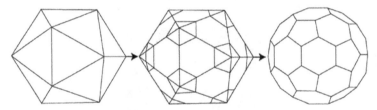

Fig. 4.3. Construction of the Archimedean truncated icosahedron from the Platonic icosahedron.

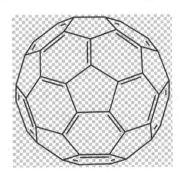

Fig. 4.4. Archimedean truncated icosahedron.

In the above-mentioned book by Wenniger, *The Models of Polyhedra* [112], the reader can find 75 different models of regular polyhedra. According to the Russian mathematician L.A. Lyusternak, who did a lot in this particular area of mathematics, *"The theory of polyhedra, in particular convex polyhedra, is one of the most fascinating chapters of geometry"*.

### 4.2.3. Stellate polyhedral

A great contribution into the development of the theory of polyhedra was introduced by Johannes Kepler (1571–1630). During the Renaissance, all Archimedean solids were "rediscovered". Finally, Kepler in 1619 in his book, *Harmonices Mundi* (*Harmony of the World*), gave an exhaustive description of the whole set of Archimedean polyhedra.

At one time, Kepler wrote a small book, *On the Snowflake*, in which he made the following remark: *"Among the Platonic solids, the cube is the very first, the beginning and the progenitor of the rest*

Fig. 4.5. Regular stellate polyhedra (Poinsot solids).

*Platonic solids, and the octahedron is its wife, because the octahedron has as many angles as the cube of the faces."*

Kepler was the first scientist, who published a complete list of the 13 Archimedean Solids and gave them the names, by which they are known today. Kepler first began to study the so-called *stellate polyhedra*, which, unlike the *Platonic* and *Archimedean solids*, are *regular non-convex polyhedra*. The French mathematician and mechanic L. Poinsot (1777–1859), whose geometric works relate to stellate polyhedra, discovered the existence of the new types of regular non-convex polyhedra (Fig. 4.5).

## 4.3. The Mystery of the Egyptian Calendar

### 4.3.1. What is a calendar?

The Russian proverb says: "Time is the eye of history." Everything that exists in the Universe, the Sun, the Earth, the stars, the planets, the known and unknown worlds, and everything that is in the Nature has a space–time dimension. Time is measured by observing periodically repetitive processes of certain duration.

Already in the ancient times, people noticed that the day always gave way to night, and the seasons passed in a strict sequence: spring came after winter, summer came after spring, and autumn came after summer. In search of the explanation of these phenomena, man drew attention to the heavenly solids: the Sun, the Moon, stars and the rigorous periodicity of their movement through the sky. These were the first observations that preceded the birth of astronomy, one of the most ancient sciences.

Astronomy is based on the movement of celestial solids, which reflects three factors: the rotation of the Earth around its axis, the rotation of the Moon around the Earth, and the movement of the Earth around the Sun. According to this, different concepts of time exist. Astronomy covers *sidereal time, solar time, local time, atomic time,* etc.

The Sun, like all other heavenly solids, is involved in the movement on the sky. In addition to daily movement, the Sun has the so-called annual movement, and the entire path of the annual movement of the Sun on the sky is called the *ecliptic*. If, for example, we notice the location of the constellations at any particular evening hour and then repeat this observation every month, then we will see a different picture of the sky. The view of the starry sky changes continuously: each season has its own pattern of evening constellations, and each such pattern is repeated a year later. Consequently, after the year has passed, the Sun returns to its former place with respect to the stars.

For convenience of the orientation in the stellar world, astronomers divided the entire sky into 88 constellations. Each of them has its own name. Among the 88 constellations, the special place in astronomy is occupied by those, through which the *ecliptic* passes. These constellations, in addition to their own names, also have a generic name, *zodiacal* (from the Greek word "zoop", meaning animal), as well as symbols (signs) well known all over the world and various allegorical images, included in *calendar systems*.

It is known that in the process of moving along the *ecliptic*, the Sun crosses 13 constellations. However, astronomers found it necessary to divide the path of the Sun not into 13, but into 12 parts, by combining the constellations Scorpio and Ophiuchus into one, under the general name Scorpio.

As it is known, a special science, called *chronology*, deals with the problems of measurement of time. The chronology is the basis of all calendar systems created by mankind. The creation of calendars in antiquity was one of the most important tasks of astronomy.

The word "calendar" comes from the Latin word *"calendarium"*, which literally meant "debt book"; in such books, the first day of each month was specified and they were called "calendas": in these days, debtors paid interest in ancient Rome.

Since ancient times in the countries of East and Southeast Asia, while drawing up calendars, great importance was attached to the periodicity of the movement of the Sun, the Moon, as well as Jupiter and Saturn, the two giant planets of the Solar System. There is reason to assume that the idea of creating the Jupiterian calendar with the celestial symbolism of the 12-year animal cycle is associated with the rotation of Jupiter around the Sun; Jupiter makes a complete rotation around the Sun in about 12 years (11.862 Earth years). On the other hand, Saturn, the second giant planet of the Solar system, makes a complete rotation around the Sun in about 30 years (29.458 Earth years). *By wanting to harmonize the cycles of motion of giant planets, the ancient astronomers came to the idea of introducing the 60-year cycle of the Solar system. During this cycle, Saturn makes two full rotations around the Sun, and Jupiter makes five rotations.*

When creating annual calendars, various astronomical phenomena were used: the change of day and night, the change of the lunar phases and the change of seasons. The use of various astronomical phenomena led to the creation of three types of calendars in different nations: *lunar*, based on the movement of the Moon, *solar*, based on the movement of the Sun, and *lunar–solar*.

### 4.3.2. The structure of the Egyptian calendar

One of the first *solar* calendars was the Egyptian calendar, created in the fourth millennium BC. Initially, the Egyptian calendar year consisted of 360 days. The year was divided into 12 months, exactly 30 days each. However, later it was discovered that such a calendar year did not correspond astronomically. Then the Egyptians added 5 more days to the calendar year; however, these additional 5 days were not the days of the months. These were five public holidays, which connected the neighboring calendar years. Thus, the Egyptian calendar year had the following structure: $365 = 12 \times 30 + 5$. Note that it is the Egyptian calendar that is the prototype of the modern calendar.

The following question arises: Why did the Egyptians divide the calendar year into 12 months? After all, there were calendars with other amount of months in one year. For example, in the Mayan calendar, a year consisted of 18 months, with 20 days in one month. This raises the next question, concerning the Egyptian calendar: Why did every month have exactly 30 days?

We can also ask some questions about the choice of the modern system of time measurement, in particular, about the choice of such time units as *hour*, *minute*, and *second*. In particular, the question arises: Why the unit of one *hour* was chosen so that it fit exactly 24 times in a day, that is, why is 1 day $= 24 = 2 \times 12$ hours? Further, why 1 *hour* $= 60$ *minutes*, and 1 *minute* $= 60$ *seconds*?

The same questions apply to the choice of measurement units of angular values, in particular: Why is the circle divided into 360°, that is, why $2\pi = 360° = 12 \times 30°$? Other questions are added to these questions, in particular: Why did astronomers consider it appropriate to assume that there are 12 *zodiacal constellations*, although in fact, the Sun crosses 13 constellations during its movement along the ecliptic? And one more "strange" question: Why did the Babylonian numeral system have a much unusual base, the number of 60? Is there any connection between these facts?

### 4.3.3. Connection of the Egyptian calendar with the numerical characteristics of the dodecahedron

By analyzing the Egyptian calendar, as well as modern systems for measuring time and angular values, we find that the four numbers are repeated with surprising consistency: 12, 30, 60 and the number $360 = 12 \times 30$ derived from them. The question arises: Is there any fundamental scientific idea that could provide a simple and logical explanation for the use of these numbers in the Egyptian calendar and modern system of time and angles measurement?

To answer this question, let's once again turn to the dodecahedron (Fig. 4.6).

Fig. 4.6. Dodecahedron.

The question arises: Did the Egyptians know the dodecahedron? Historians of mathematics recognize that the ancient Egyptians possessed information about the regular polyhedra. But did they know all the five regular polyhedra, in particular, dodecahedron and icosahedron as the most complex of them?

The ancient Greek philosopher and mathematician Proclus attributes the construction of regular polyhedra to Pythagoras. But after all, many mathematical theorems and results (in particular, the *Pythagoras theorem*) Pythagoras borrowed from the ancient Egyptians during his very long "business trip" to Egypt (according to some information, Pythagoras lived in Egypt for 22 years!). Therefore, we can assume that the knowledge about the regular polyhedra Pythagoras, perhaps, also borrowed from the ancient Egyptians (and possibly from the ancient Babylonians because according to legend, Pythagoras lived in ancient Babylon for 12 years). But there are other, more compelling evidence that the Egyptians possessed information about all the five regular polyhedra. In particular, in the British Museum, one could see the dice of the era of the Ptolemies, which has the shape of *icosahedron*, that is, the *Platonic solid*, dual to *dodecahedron*. All these facts give us the right to put forward the hypothesis that the *dodecahedron* was known to the Egyptians. If this is so, then the much harmonious system follows from this hypothesis, allowing one to give the simple explanation of the origin of the Egyptian calendar, and at the same time, the origin of the modern system for measuring time intervals and geometric angles.

Earlier, we established that the dodecahedron has 12 faces, 30 edges and 60 flat angles on its surface (Table 4.1). If we proceed from the hypothesis that the Egyptians knew the dodecahedron and its numerical parameters, then it is not clear what so much surprised the Egyptians when they discovered that the cycles of the Solar system, namely, the 12-year cycle of Jupiter, the 30-year cycle of Saturn and, finally, the 60-year cycle of the Solar system. Thus, there is a profound mathematical connection between such a perfect spatial figure as the dodecahedron (Fig. 4.6) and the Solar system!

Such a conclusion was made by the Egyptian astronomers. This led them to the fact that the dodecahedron was adopted as the "main figure", which symbolized the *Harmony of the Universe*. And then the Egyptian astronomers decided that the Egyptian calendar correspond to the numerical parameters of the dodecahedron!

Because, according to the idea of the Egyptian astronomers, the movement of the Sun along the *ecliptic* had a strictly circular character, then, by choosing the 12 zodiacal signs, the arc distance between which was equal exactly to 30°, the Egyptians surprisingly beautifully coordinated the annual movement of the Sun along the ecliptic with the structure of their calendar year: one month corresponded to the movement of the Sun along the ecliptic between two adjacent signs of the Zodiac! Moreover, by moving the Sun on one degree corresponded to one day in the Egyptian calendar year! In this case, the ecliptic was automatically divided into 360°.

Later, the dodecahedral idea of the Egyptian astronomers was borrowed by modern astronomers, who decided that all their main astronomical systems (in particular, the time and angle measurement units: *hour, minute, second*) should correspond to the *dodecahedron*. This led modern astronomers to the fact that the dodecahedron was adopted as the "main geometric figure", which symbolized the *Harmony of the Universe*.

Thus, by choosing the *dodecahedron* as the main "harmonic" geometric figure of the Universe, and strictly following its numerical characteristics, the Egyptian astronomers created the Egyptian calendar, a prototype of the modern calendar, and later, the astronomers created the time and geometric angle measurement systems, based on the numerical characteristics of the dodecahedron. *It should be emphasized that the Egyptian calendar and the modern systems for measuring time and angular values were fully consistent with the ancient "Theory of Harmony", based on the golden ratio because it is this proportion that underlies the dodecahedron.*

Figure 4.7 presents a modern calendar made in the form of the *dodecahedron*. In this calendar, each of the 12 faces of the dodecahedron corresponds to certain calendar month.

Fig. 4.7.  Modern calendar in the form of dodecahedron.

These are the surprising conclusions from the comparison of the *dodecahedron* to the *Solar system*. But the main proportion of dodecahedron is the *golden proportion*! Therefore, if our hypothesis is correct (let someone try to refute it), then it follows from these consideration that for many millennia, humanity lives under the sign of the *golden section*! And every time when we look at the dial of our watch (Fig. 4.8), which is also constructed on the use of the numerical characteristics of the dodecahedron, we touch the main "Mystery of the Universe", the *golden section*, without knowing this!

### 4.3.4.  About the Mayan calendar

It is known that the calendar year in the Mayan calendar had the following structure: 1 year $= 360 + 5 = 20 \times 18 + 5$ days, whence it follows that the Mayan year was divided into 18 months, 20 days each. The numbers 20 and 360 were used by the Mayans as the "*nodal numbers*" of their numeral system. However, in its structure, the Mayan calendar year was similar to the structure of the Egyptian calendar year: 1 year $= 360 + 5 = 12 \times 30 + 5$ days, where the numbers

Fig. 4.8. Men's wrist watch.

Fig. 4.9. Icosahedron.

12 and 30 are the numbers of the dodecahedron. But, what is the number 20 in the Mayan calendar? We now turn to the *icosahedron* (Fig. 4.9).

The icosahedron (Fig. 4.9) has 20 faces (see Table 4.1). Thus, the Mayans undoubtedly used this particular numerical characteristic of the icosahedron both in their calendar (dividing the year by 20 months) and in their numeral system (by choosing the numbers 20 and 360 as the "nodal" numbers of their numeral system).

## 4.4. Dodecahedral–Icosahedral Doctrine

### 4.4.1. The sources of the doctrine

According to the commentator of the last edition of Plato's works, *"all cosmic proportionality of his works is based on the principle of the golden section, or harmonic proportion."* As it is mentioned, Plato's cosmology is based on regular polyhedra, called the *Platonic solids*. The idea of the "through" Harmony of the Universe has always been associated with its embodiment in these five regular polyhedra.

The fact that the main "cosmic" figure, the *dodecahedron*, which symbolized the body of the world and the universal soul, is based on the *golden section*, gives the latter a special charm, as the main proportion of the Universe. Plato's cosmology became the basis of the so-called *icosahedral–dodecahedral doctrine*, which since the ancient time has been passing through the whole human science. The essence of this doctrine consists in the fact that *dodecahedron* and *icosahedron* are typical forms of Nature in all its manifestations by starting with the *cosmos* and ending with the *microworld*.

### 4.4.2. Earth shape

The best minds since the ancient times sought *Harmony* in the trinity: *Earth, Man, Cosmos*, and the idea that some simple mathematical relationships lie at the core of everything has firmly taken root in the ancient and modern science.

Even Plato (and according to other data, Socrates) stated: *"The Earth, if we look at it from above, looks like a ball stitched from 12 pieces of leather"*. All Pythagoreans represented our planet so. This hypothesis has found further scientific development in the works of physicists, mathematicians and geologists. Thus, the French geologist de Beamon (1798–1874) and the famous mathematician Poincaré (1854–1912) believed that the Earth is similar in its shape to the *deformed dodecahedron*.

Practical use of the hypothesis *"Earth is a growing crystal"* to explain the processes taking place not only in the subsoil and on the surface of the planet, but also influencing the changes of the living world and even the development of civilizations was undertaken by Soviet scientists, Goncharov, Makarov, and Morozov. In their opinion, the force field of this growing crystal causes the icosahedral–dodecahedral structure of the Earth. These polyhedra are inscribed into each other. Projections of icosahedron and dodecahedron appear on the surface of the Earth.

The presentation about the Earth as a huge growing crystal is part of the scientific ideas that began to develop intensively at the end of the 20th century. According to the increasingly attractive points of view of scientists, everything in the Universe is either a crystal or seeks to adopt an ordered crystalline structure. For us, the most important thing in these studies is the fact that in these amazing models of the structure of the Earth, *dodecahedron* and *icosahedron*, the two *Platonic solids*, which in the ancient science expressed the *Harmony of the Universe*, play a main role.

## 4.5. Johannes Kepler: From "Mystery" to "Harmony"

### 4.5.1. Misterium Cosmographicum

Among the patriarchs of modern European science, there is no figure more mysterious than Johannes Kepler (1571–1630). First of all, Kepler was a professional astrologer, a dreamer and science fiction writer, whose style of thinking was unacceptable both for the creators of classical science, including Galileo and Newton, and for its historians, at least for historians of the classical formation. On the other hand, he is the greatest scientist, who introduced many new fundamental concepts into modern science (Fig. 4.10).

The authors of the *Historical Vocabulary of Philosophy* attribute to Kepler a modern, that is, mechanistic understanding of *strength*. It turns out that he introduced the notion of *inertia*, which distinguishes modern physics from all the previous ones, and at the same time, the physical concept of *energy*, not to mention the fact that he is the author of the *first quantitative laws of astronomy*.

Fig. 4.10.  Johannes Kepler (1571–1630).

Kepler is the founder of the *physics of the sky*. This remarkable phrase is included in the subtitle of his main work: "*New astronomy based on causes, or physics of the sky.*"

Johannes Kepler was born in 1571 in a poor Protestant family. In 1591, he entered the Tubingen Academy, where he received a good mathematical education. It was there that the future great astronomer had acquainted himself with Nicholas Copernicus's *heliocentric system* of the world.

After graduating from the Academy, Kepler received a master's degree and then was sent as a mathematics teacher to the grammar school in Graz (Austria). His first astronomical essay was a small book with the following title: "*The forerunner of cosmographic research, containing the mystery of the Universe regarding the miraculous proportions between the celestial circles and the true causes of the number and size of the celestial spheres, as well as periodic movements outlined with the help of five regular bodies by Johannes Kepler from Württemberg, a mathematician from the glorious province of Styria.*" He himself called this book, published in 1597, *Misterium Cosmographicum* (*The Mystery of Cosmography*).

By reading Kepler's first essay *Misterium Cosmographicum*, we never get tired of being surprised by his imagination. The deep conviction that *Harmony* of the world exists left an imprint on Kepler's whole mindset. The goal of his research, outlined in his *Misterium Cosmographicum*, was formulated by Kepler as follows:

> "*Dear reader! In this book, I set out to prove that the all-good and almighty God in the creation of our moving world and in the arrangement of celestial orbits chose five regular solids, which since the times of Pythagoras and Plato to this day have gained such loud glory, as the basis as well as the relationship between movements, chosen in accordance with the nature of regular solids. The essence of three things — why they are arranged in such the way and not otherwise — especially interested me, namely, the number, dimensions and movements of the celestial orbits.*"

The aim was to reveal the mystery of cosmography meant, according to Kepler, to answer the question that he had set himself for the first time in the history of astronomy. It was in the book *Misterium Cosmographicum* that Kepler revealed this secret. Its essence, according to Kepler, is as follows:

> "*The Earth (the orbit of the Earth) is the measure of all orbits. We will describe the dodecahedron around it. The sphere, described around the dodecahedron, is the sphere of Mars. We describe a tetrahedron around the sphere of Mars. The sphere, described around the tetrahedron, is the sphere of Jupiter. Around the sphere of Jupiter we describe the cube. The sphere, described around the tetrahedron, is the sphere of Saturn. We put an icosahedron into the sphere of the Earth. The sphere, inscribed in it, is the sphere of Venus. Let's put the octahedron in the sphere of Venus. The sphere, inscribed in it, is the sphere of Mercury.*"

The famous Kepler *Cosmic Cup* (Fig. 4.11), which sets crystalic spheres into Platonic solids, embodies this model in matter. The most precious heritage of the ancient geometry (Platonic solids) had been putting in, finally, to Pythagorean astronomy.

Of course, the creation of the *Cosmic Cup* was a great success for the young astronomer; this scientific result brought for Kepler a scientific glory. The *Cosmic Cup* provided him with access to the invaluable observational data collected in the *Castle of Heaven* by Tycho Brahe. Kepler assisted the great astronomer in the study of Mars. The *Cosmic Cup* allowed Kepler to come to the important conclusion about the "mystery of the Universe": The Universe was arranged on the basis of a single geometric principle!

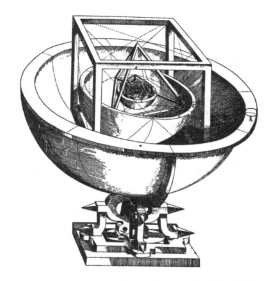

Fig. 4.11. *Cosmic Cup* as Kepler's model of the Solar system.

But the joy proved to be premature. With all his exhilaration, Kepler was endowed with all the qualities, inherent to a scientist. Kepler understood that theory should be consistent with the results of observations. While restraining back the delight, which seized Kepler at the thought of his wonderful discovery, Kepler began checking his model.

The unified geometrical principle, from the Cosmic Cup, allowed Kepler to answer the two questions posed by him:

(1) explain the number of planets known at that time (with the help of the five "Platonic solids", the six spheres can be constructed) that led to the conclusion about the existence of the six planets;

(2) answer the question about the distance between the planets.

The answer to the third question (about the motion of the planets) turned out to be the most difficult, and it was received by Kepler many years later.

We will not dwell on all the scientific discoveries and achievements of Johannes Kepler. They are widely known. It is enough to mention only about Kepler's *three famous astronomical laws*, which immortalized his name.

But with Kepler's death (in 1630), he was forgotten in the science of the *golden section*, which Kepler considered to be one of the *"treasures of geometry"*, comparable to the *Pythagoras theorem*. And this strange oblivion lasted for almost two centuries. The interest in the *golden section* and *Fibonacci numbers* was revived only in the 19th century in the works of the French mathematicians, Lucas and Binet (see Chapter 2).

## 4.6. Icosahedron as the Main Geometric Object of Mathematics

### 4.6.1. Felix Klein

As mentioned, *dodecahedron* and *icosahedron* occupy a special place among the five *Platonic solids*. These two *Platonic solids* are directly connected with the *pentacle* and through it with the *golden ratio*. *Dodecahedron* and *icosahedron* underlie the so-called *dodecahedral–icosahedral doctrine*, which permeates the entire history of human culture by starting from Pythagoras and Plato. And probably, it cannot be accidental that this doctrine was developed in the works of the outstanding German mathematician, Felix Klein (1849–1925) (Fig. 4.12).

Felix Klein was born in Dusseldorf in 1849. In 1865, he entered the University of Bonn. In 1872, Klein worked in Erlangen, since

Fig. 4.12. Felix Klein (1849–1925).

1875, he was a Professor at Higher Technical School in Munich, and since 1880 he worked as a Professor at Leipzig University. In 1886, he moved to Göttingen, where he headed the Mathematics Institute at the University of Göttingen; this Mathematics Institute during the first quarter of the 20th century was recognized as the world's mathematical center.

Klein's main works were devoted to the *non-Euclidean geometry*, the *theory of continuous groups*, the theory of *algebraic equations*, the theory of *elliptic functions*, and the theory of *automorphic functions*. Klein outlined his ideas in the field of geometry in his work *A Comparative Consideration of New Geometric Studies* (1872), known as the *Erlangen Program*.

### 4.6.2. Symmetry groups of regular polyhedra

According to Klein, each geometry is a theory of invariants of a special group of transformations. By expanding or narrowing the group, we can move from one type of geometry to another. The *Euclidean geometry* is the science of invariants of a *metric group*, the *projective geometry* is the science of invariants of a *projective group*. The classification of transformation groups gives us a classification of geometries. A significant Klein's achievement is the proof of the consistency of the *non-Euclidean geometry*.

By exploring discrete groups, Klein had considered the symmetry groups of the regular polyhedra. Let's consider these concepts for specific examples. Let's start with the *plane of symmetry*. The plane of symmetry in Nature appears very often. The outstanding Russian crystallographer G.V. Wulf (1863–1925) called the plane of symmetry *"the main key element of symmetry"*.

It is easy to make sure that a square has four *planes of symmetry*, that is, 4*P*, and a rectangle has only two *planes of symmetry*, that is, 2*P*. In the same way, the symmetry planes can be found for more complex geometric objects. It is easy to verify that a *cube* has nine planes of symmetry (Fig. 4.13), that is, 9*P*.

Now let's consider the second type of the symmetry elements called the *axis of symmetry*. The most graphic example of the concept of the *axis of symmetry* can be a faceted glass in a cup holder; the

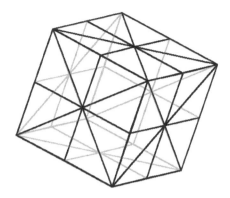

Fig. 4.13. The nine symmetry planes of the cube.

shape of the faceted glass exactly repeats the shape of the cup holder. By having removed a glass from such a cup holder and by having put it back in a turned position, we essentially perform the operations of the *self-combining ones*. The number of the various *self-combining ones*, which can be performed around the given axis, is called its *order*.

Usually, the *axis of symmetry* is denoted by the letter $L$, and its *order* by a small number, following this letter. For example, $L_3$ denotes the *axis of symmetry of the third order*. It is clear that an equilateral triangle has the *axis of symmetry* $L_3$, a square $L_4$, a regular pentagon $L_5$, and the circle has the *axis of symmetry* of infinite order $L_\infty$.

As an example for demonstrating the *axes of symmetry*, let's consider the *cube* again (Fig. 4.13). Perpendicular to each pair of cube faces, the *quadruple axis of symmetry* passes through the centers of the squares (Fig. 4.13). This means that the cube has three axes of symmetry of the fourth order, that is, $3L_4$. The cube has eight vertices. Through each opposite pair of vertices, there passes a *triple axis of symmetry*, which coincides with the solid diagonal of the *cube* (Fig. 4.13). This means that the cube has four axes of symmetry of the third order, that is, $4L_3$. The *cube* has 12 edges. Through the midpoints of each pair of edges, parallel to the diagonals of the faces, the *double axis of symmetry* passes (Fig. 4.13). This means that the *cube* has six *axes of symmetry of the second order*, that is,

$6L_2$. Therefore, the full characteristic of the axes of symmetry of the *cube* is as follows: $3L_4$; $4L_3$; $6L_2$.

There are geometric solids with symmetry axes of the infinite order $L_\infty$. The so-called *rotation solids* (a cylinder, a cone, etc.) have such axes of symmetry. Any diameter of the ball has the axis of symmetry $L_\infty$. This means that the ball has an infinite set of the symmetry axes of the infinite order, that is, $\infty L_\infty$.

Let's consider now one more element of symmetry called the *center of symmetry C*. The *cube* is a vivid example of the geometric figure, which has the center of symmetry.

Usually, to characterize the symmetry of any geometric object, a complete set of symmetry elements is given. For example, the symmetry group of snowflakes has the form $L_6 6P$. This means that the snowflake has one symmetry axis of the order 6 $L_6$, that is, it can *self-align* six times, when rotating around the axis, and six planes of symmetry. The symmetry group of the daisy flower, having 24 petals, has the form $L_{24}\, 24P$, that is, this flower has one symmetry axis of the 24th order and 24 planes of symmetry $P$. By summing up all the symmetry elements of the *cube*, we get the next symmetry group: $3L_4\, 4L_3\, 6L_2\, 9P\, C$.

As we already know, the *cube* is one of the five *Platonic solids*. Table 4.3 shows the symmetry groups of all the *Platonic solids*.

The analysis of the symmetries of the *Platonic solids*, given in Table 4.3, shows that the symmetry groups of *cube* and *octahedron*, as well as *dodecahedron* and *icosahedron*, coincide. This is due to the fact that the *dodecahedron* is dual to the *icosahedron*, and the *cube* is dual to the *octahedron*.

Table 4.3. Symmetry groups of Platonic solids.

| Polyhedron | Shape of faces | Symmetry group |
|---|---|---|
| Tetrahedron | Equilateral trinagles | $4L_3\, 3L_2\, 6P$ |
| Hexahedron ( cube) | Squares | $3L_4\, 4L_3\, 6L_2\, 9P\, C$ |
| Octahedron | Equilateral trinagles | $3L_4\, 4L_3\, 6L_2\, 9P\, C$ |
| Icosahedron | Equilateral trinagles | $6L_5\, 10L_3\, 15L_2\, 15P\, C$ |
| Dodecahedron | Reglar pentagons | $6L_5\, 10L_3\, 15L_2\, 15P\, C$ |

### 4.6.3. The role of the icosahedron in the development of modern science

In addition to the *Erlangen program* and other outstanding mathematical achievements, Felix Klein's genius was also manifested in the fact that more than 100 years ago, he was able to predict the outstanding role of the *Platonic solids*, in particular the *icosahedron*, in the future development of science, in particular, mathematics. In 1884, Felix Klein published the book *Lectures on the Icosahedron and the Solution of the Fifth-Degree Equations* [113], devoted to the geometric theory of the *icosahedron*. The first part of the book defines the place of the icosahedron in mathematics. According to Klein, the fabric of mathematics has been widely and freely scattered to sheets of individual theories. But there are geometric objects, in which several sheets converge into peculiar branch points. Their geometry links these sheets and allows covering the general mathematical sense of different theories. According to Klein, the *icosahedron* is such a geometric object. Klein treats the *icosahedron* as a geometric object, from which the branches of the five mathematical theories diverge: *geometry, Galois theory, group theory, theory of invariants and differential equations*. Thus, Klein's main idea is extremely simple: "*every unique geometric object is somehow connected with the properties of the icosahedron*".

What is the significance of the ideas of the outstanding mathematician from the point of view of the *Theory of Harmony*? First of all, the *icosahedron*, based on the *golden section*, was chosen as a geometric object, combining the *master sheets* of mathematics. From here, naturally, we can conclude that it is the *golden section* that is the main geometric idea, which, according to Klein, can unite all mathematics.

Klein's contemporaries failed to properly understand and appreciate the revolutionary character of Klein's "icosahedral" idea. Its significance was understood exactly 100 years later, that is, only in 1984, when the Israeli physicist Dan Shechtman published a note, confirming the existence of special alloys (called *quasicrystals*) with the so-called *icosahedral symmetry*, that is, symmetry of the order 5, which is strictly prohibited in classical crystallography. In 2011, the author of this discovery, Dan Shechtman, was awarded the Nobel Prize in Chemistry.

Thus, back in the 19th century, Felix Klein's ingenious intuition led him to the very important scientific fact that the *icosahedron*, one of the most beautiful *Platonic solids*, is the main geometric figure of mathematics. Thus, Klein in the 19th century breathed new life into the development of the *"dodecahedral–icosahedral idea"* about the structure of the Universe, the followers of which were great scientists and philosophers: Plato, who built his cosmology on the basis of regular polyhedra, Euclid, who devoted his *Elements* to the theory of the *Platonic solids*, Johannes Kepler, who used the *Platonic solids* when creating his Cosmic Cup, the original geometric model of the Solar system.

### 4.7.  Usage of the Regular Polyhedra in the Fine Art

In 2002, the scientific journal, *Energy*, published a series of articles, entitled "Art and science: about polyhedra in general and a truncated icosahedron in particular" [114], written by the Russian researcher Ye.A. Katz. The article contains a lot of interesting information about the use of the regular polyhedra in fine art. The materials in these articles are used by the authors when writing this section.

### 4.7.1.  Leonardo da Vinci image techniques for regular polyhedra

Many authors note the original ways of spatial representation of *icosahedron, dodecahedron, truncated icosahedron*, and other polyhedra, authored by Leonardo da Vinci, and reproduce these beautiful images from the illustrated Leonardo book of his contemporary, Franciscan monk and mathematician Luca Pacioli (1445–1514) *Divine Proportion*, published in 1509. Apparently, it is impossible to consider Leonardo's involvement in the study of these perfect geometric figures as an accident. Moreover, it is deeply symbolic. The titan of the Renaissance, painter, sculptor, scientist and inventor Leonardo da Vinci (1452–1519) is a symbol of the continuity of art and science, and therefore, his interest in such beautiful, highly symmetrical objects like convex polyhedra is very

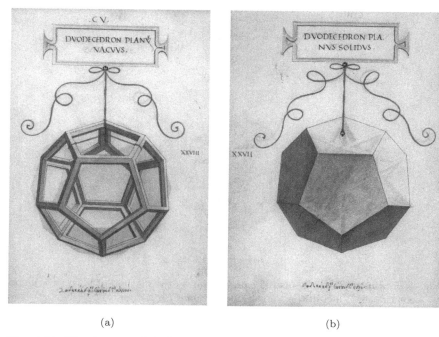

(a)                                              (b)

Fig. 4.14. The images of dodecahedron by (a) the rigid ribs method and (b) the compact faces method, performed by Leonardo da Vinci in the book *Divine Proportion* by Luca Pacioli (1509).

logical. Figure 4.14 presents the images of *dodecahedron*, made by Leonardo da Vinci for the book *Divine Proportion* by Luca Pacioli by using two methods: the method of the *rigid ribs* (Fig. 4.14(a)) and the method of the *compact faces* (Fig. 4.14(b)).

A comparison these methods on the example of *dodecahedron* convincingly shows the advantage of the *method of rigid ribs*. The essence of the method of the *rigid ribs* is that the faces of *dodecahedron* are depicted as "*empty*", that is, *not compact*. Strictly speaking, the faces are not depicted at all, because they exist only in our imagination. However, the *ribs* of the *dodecahedron* are shown not by geometric lines (which, as is well known, have neither width nor thickness), but by the *rigid* three-dimensional segments.

Such a technique of depicting polyhedra allows the viewer, firstly, to accurately determine which ribs belong to the front and which ribs to the rear faces of the polyhedron (what is almost impossible when

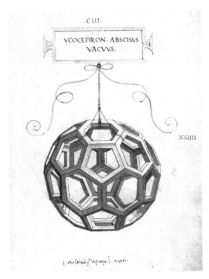

Fig. 4.15. Image of the truncated icosahedron by the method of hard ribs, made by Leonardo da Vinci in the Pacioli book *Divine Proportion*.

depicting ribs by geometric lines) and, secondly, to see as if through a geometric body and to feel it in perspective, which is lost, when we use the technique of the compact faces.

Figure 4.15 shows the image of the *truncated icosahedron*, made by Leonardo da Vinci, according to the *hard ribs method* for the book by Luca Pacioli. An engraving depicting the *truncated icosahedron* (Fig. 4.15) is prefaced by Leonardo by the inscription in Latin "*Ycocedron Abscisus Vacuus*" (truncated icosahedron). The term *Vacuus* just means that the faces of the *truncated icosahedron* are depicted as "*empty*".

Note that Johannes Kepler used the method of the rigid ribs when depicting polyhedrons that make up his Cosmic Cup (Fig. 4.11).

### 4.7.2. Polyhedra by Luca Pacioli

Pacioli's book, for which Leonardo made 60 illustrations of various polyhedra, had a great influence on the development of the geometry of that time, in particular, on the stereometry of polyhedra.

As it is known, Pacioli was also one of the largest European algebraists of the 15th century and, no less important, he invented the principle of the so-called *double record*, which is currently used in all accounting systems without exception. So, he can be safely called the *"father of modern accounting"*.

However, the rather mysterious and controversial personality of Pacioli to this day causes fierce disputes between historians of science. It is authentically known that Luca Pacioli was born in 1445 in the Italian town of Borgo San Sepolcro. As a child, he studied in the studio of the famous painter and mathematician Piero della Francesca (1415–1492), and then at the University of Bologna, which was one of the best in Europe in the 15th and 16th centuries (at various times, his students were, for example, Copernicus and Dürer).

In 1472, Pacioli under the name Fra Luca di Borgo-San Sepolcro returned to his hometown and began to work on the most famous of his writings, the book *Summary of Arithmetic, Geometry, Proportions and Proportionality*, published in Venice in 1494. In 1496, he was invited to lecture mathematics in Milan University. In Milan, he met Leonardo da Vinci. Leonardo, after reading Pacioli's Summade arithmetica, abandoned work on his own book on geometry and began to prepare illustrations to Pacioli's new book, *Divine Proportion*.

Some researchers accuse the author of *Divine Proportion* of plagiarism from the unpublished manuscripts belonging to Pacioli's teacher Piero della Francesca. Others, on the contrary, protect Pacioli from these accusations. In general, this is the dark history.

But Pacioli's appearance is known to us due to the painting "Luca Pacioli" (Fig. 4.16) by the painter Jacopo de Barbari (1440–1515). Barbary's work is excellent in all respects and, above all, in the transfer of the depicted personalities. Every detail of the composition in Barbary's painting is full of deep meaning. The artist shows a deep understanding of the relationship between art and science, which was so characteristic of the Renaissance masters. Pacioli in a Franciscan monk's cassock is depicted standing at the table with geometric tools and books (in the lower right corner of the picture, we can see a model of the *dodecahedron*).

Fig. 4.16.  The painting "Luca Pacioli" by artist Jacopo de Barbari.

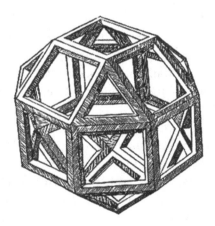

Fig. 4.17.  Rhombic cuboctahedron or polyhedron by Luca Pacioli.

The attention of Pacioli and the handsome young man, standing to the right and slightly behind Pacioli, is focused on the study of the polyhedron, the model of which we see in the upper left corner of the composition. The choice of this polyhedron is not accidental: it is a *rhombic cuboctahedron* or *Pacioli's polyhedron* (Fig. 4.17). According to the modern mathematician and artist George Hart, Pacioli himself chose this polyhedron for the painting because Pacioli was especially proud of this geometric discovery.

### 4.7.3. Albrecht Dürer

The personality of the young man, standing next to Pacioli in the painting of Jacopo de Barbari (Fig. 4.16) is still a topic of controversial debates between art historians: some of them believe that this is Barbari himself; others identify him as Albrecht Dürer (1471–1528), the artist and graphic artist, the greatest representative of the German Renaissance. This assumption is at least debatable. But, and this is much more important in our context, another is known for certain. Dürer was struck by the artistic style of Barbari, who built his compositions on the basis of an in-depth study of the system of proportions, that is, a strictly defined ratio of the parts, depicted to each other (Fig. 4.18).

Dürer began to study the laws of perspective, he dreamed to meet with famous Italian masters, learn from them, and compete with them. To this end, in 1505, Dürer embarks on a journey through Italy. It is not known exactly who his teachers were in the school of perspective (among other names, the names of Luca Pacioli and Piero

Fig. 4.18. Albrecht Dürer (1471–1528).

Francesca are named), but studying at this school, Dürer continued throughout his life.

Three years before his death, in 1525, a 54-year-old master, the author of more than 60 paintings and several hundred engravings, hastened to share with his descendants the secrets of perspective, accumulated by him during his life. He published a treatise *Guide to Measurement* and then two more: *A Manual for Strengthening Cities* (1527) and *Four Books on Proportions* (1528).

Dürer's books are a serious scientific contribution to the theory of perspective and stereometry of polyhedra. He was the first to describe several Archimedean solids, unknown at that time, and also develop and first publish, in his book, models of flat sweeps of various polyhedra, including the truncated icosahedron. Nowadays, such sweeps, from which three-dimensional models of polyhedrons are assembled, are widely used to study the elementary forms of crystals, the structure of molecules (fullerenes, for example), viruses, etc.

All the fine art of that era permeated by thirst for knowledge, and scientists–naturalists became the largest representatives of the fine art of that era. The idea of the unity of artistic inspiration and mathematical theory is also reflected in the famous 1514 engraving by Dürer's "*Melancholia*" (Fig. 4.19). The presence of a polyhedron (most likely a truncated rhombohedron) on the engraving, of course, is not accidental.

Hundreds of pages are written by art historians in an attempt to explain the meanings of the symbols used by Dürer in engraving "*Melancholia*". One of them, E. Panovski, writes:

> "*Dürer presented "Melancholy" as one of the four temperaments and as geometry, one of the seven free arts. He embodies in "Melancholy" the type of Renaissance artist, who appreciates practical skill, does not avoid mathematical theory, and who, feeling involved in divine inspiration, simultaneously suffers from all human imperfection and limitedness. So this is, in some sense, the spiritual self-portrait of Durer.*"

### 4.7.4.  Piero della Francesca

Many artists from different eras and countries experienced a constant interest in the study and depiction of polyhedra. The peak of

Fig. 4.19. Albrecht Dürer Melancholy.

this interest falls, of course, on the Renaissance. By studying the phenomena of Nature, the Renaissance artists sought to find methods of the image of Nature phenomena by using the experience of science. The doctrines about perspective and proportions, based on mathematics, optics, anatomy, became the basis of a new art. They allowed the artists to recreate three-dimensional space on the surface to achieve the impression of relief of objects. For some Renaissance masters, the polyhedra were simply a convenient model for training perspective skills. Others admired their symmetry and concise beauty. Third set of artists, following Plato, were attracted by their philosophical and mystical symbols.

To the list of the greatest masters of the Renaissance, who often depicted and deeply studied the geometry of polyhedra (except the above-mentioned Leonardo da Vinci and Albrecht Dürer) should be added the name of Piero della Francesca (about 1420–1492). About the life and personality of Piero della Francesca, a brilliant artist, a serious art theorist and an outstanding geometer, little reliable information exists. It is known that he was born in the family of an artisan in the small town of Borgo San Sepolcro in Umbria, he studied in Florence, then worked in a number of Italian cities, including Rome. Piero della Francesca's works of art went beyond the local painting schools and influenced the art of the Italian Renaissance on the whole. However, few know that Piero della Francesca was an outstanding mathematician, who made, in particular, a significant contribution to the theory of polyhedra. He was an unquestioned authority in geometry and the science of perspective. Unfortunately, after his death (1492), the name of the Francesca scientist was forgotten for a long time. This happened largely due to the fact that, apparently, immediately after the death of the artist, Luca Pacioli, who was a student of Francesca, published most of Francesca's works in his book *Divine Proportion* (without reference to the authorship of Francesca). Fortunately, at the beginning of the 20th century, the originals of the three mathematical manuscripts of Francesca were found (they are now in the Vatican Library). After five centuries of oblivion, the fame of Piero della Francesca as the great mathematician of the Renaissance returned to him. Currently, it is known for certain that it was Piero della Francesca, the first of the Renaissance scientists, who rediscovered (not knowing, of course, that this was already done by Archimedes) and described Archimedes' solids in detail, in particular the five truncated Platonic solids: truncated tetrahedron, octahedron, dodecahedron and, most importantly, the truncated icosahedron. In his manuscript *On the Five Regular Solids* (*Libellus de quinque corporibus regularibus*), dated 1480, the oldest extant image of a truncated icosahedron was described.

Figure 4.20 provides a brief analysis of the artistic creativity of Piero della Francesca as the famous Renaissance artist on the example of his widely known painting *The Baptism of Christ*.

The painting *The Baptism of Christ* (1455), stored in the London Gallery (Fig. 4.20), is one of the clearest examples of his artistic style.

Fig. 4.20. Francesko's painting *The Baptism of Christ* and its harmonic analysis based on the golden section.

In the center of the picture is Christ. His ankles deep in the waters of the river, his hands are folded in the gesture of Catholic prayer. John the Baptist stands next to Christ; he pours water from a saucer onto the head of Christ (baptism by dousing). A dove is hovering over Christ's head like the symbol of the Holy Spirit.

### 4.7.5. Intarsia art

In the late 15th and early 16th centuries, the intarsia art was very popular in Northern Italy; this was a special type of inlay, a mosaic, made up of thousands of small pieces of various wood species. An example of this applied art with the image of polyhedra is shown in Fig. 4.21. This mosaic was created by Fra Giovanni da Verona (1457–1525) for the church of Santa Maria in Organo in Verona, approximately in 1520. The image of half-open shutters creates a three-dimensional effect on a flat mosaic, which is enhanced with the image of polyhedra (including the truncated icosahedron) in the technique of the hard ribs developed by Leonardo da Vinci.

Fig. 4.21. Fra Jovani da Verona mosaic created for the church of Santa Maria in Organo in Verona.

### 4.7.6. The sacrament of the last supper by Salvador Dali

It is impossible to resist the enthusiasm while looking at the *dodecahedron*, drawn by the famous 20th century artist, Salvador Dali (1904–1989) (Fig. 4.22).

As well known, Salvador Dali is a Spanish artist. He was one of the most famous representatives of surrealism. As an excellent draftsman and painter, Dali created images that looked like nightmarish visions, which he himself called *"painted snapshots of dreams"*. Some of the most frequently repeated images, such as watches losing their shape under the rays of the Sun, have become a kind of Dali's trademark.

In 1940, Dali left to the United States, where he led a reclusive life. In 1955, Dali returned to Spain. He was also engaged in sculpture, design of books, his authorship belonged to jewelry, decorations for the theater and cinema. However, some of his works were controversial enough and many critics believed that after the 1930s, he did not create anything worthy of attention, although his famous

Fig. 4.22. Salvador Dali (1904–1989).

Fig. 4.23. Salvador Dali — The Sacrament of the Last Supper.

painting *The Sacrament of the Last Supper* (1955) (Fig. 4.23) refutes this opinion.

According to traditional symbolism, the painting *The Sacrament of the Last Supper* depicts 13 figures (Christ and his 12 Apostles) sitting around the table. Christ is the central figure placed directly on the horizon line. The source of sunlight directly behind him makes Christ's figure as the focus of the painting. Christ points upward and directs the viewer's attention to the dominating transparent torso with arms stretched outward spanning the width of the picture plane. The scene's setting is within a transparent *dodecahedron*, which expressed spiritual harmony, moral purity and greatness.

### 4.7.7.  Creativity of Maurits Escher

Escher Maurits Cornelis (1898–1972) is the famous Dutch graphic artist. He was born on June 17, 1898, in Leeuwarden (Holland) in the family of a hydraulic engineer. In 1919, he entered the School of Architecture and Decorative Arts in Haarlem, but soon left the architecture for the sake of graphic arts. Until 1937, he traveled extensively in Europe, made sketches, paying particular attention to the deceptive, ambiguous elements of the landscape (Fig. 4.24).

Escher's work is highly esteemed by scientists, in particular, by mathematicians and crystallographers. M.P. Shaskolskaya, one of the founders of the Soviet school of crystallography and a student of

Fig. 4.24.  Escher Maurits Cornelis (1898–1972).

academician A.V. Shubnikov, in his book *Essays About the Properties of Crystals* writes as follows:

> "*Each crystallographic congress is usually accompanied by exhibitions: crystallographic equipment, books, photographs, the best artificially grown crystals. And at the Crystallographic Congress in Cambridge (1960), an exhibition of paintings by the Dutch painter Maurits Escher became an event.*
>
> *The artist himself, an elderly man with a narrow swarthy face, lively eyes and a small beard, was present as a delegate to the Congress, gave explanations to his drawings and told about them in his speech at the Congress. No, Escher was not a crystallographer, he was an artist and graphic artist, who graduated from the Harlem School of Architecture in 1922 and then continued his artistic education in Spain and Italy, and became known to the world for many art exhibitions. And now his drawings attracted the attention of crystallographers.*
>
> *The artist and crystallography? What is common between them? But the fact is that Maurits Escher in his drawings, as it were, discovered and intuitively illustrated the laws of the combination of symmetry elements, i.e., those laws that rule over the crystals, determining both their external form, their atomic structure, and their physical properties. Escher likes periodic drawings, making mosaic patterns of repeating shapes. He inscribes one image into another, so that the identical figures are periodically repeated, and there are no empty spaces between them.*"

But after all, this is the law according to which particles are placed in the structure of a crystal, the law of the closest packing: periodic repetition of identical groups of particles without gaps and violations.

### 4.7.8. Matjuska Teja Krasek's World

Serbian painter Matjuska Teja Krasek received a bachelor's degree from the College of Visual Arts (Ljubljana, Slovenia) and is a freelance artist (Fig. 4.25). She lives and works in Ljubljana. Her theoretical and practical work focuses on symmetry as a cohesive concept between art and science. The artworks by Matjuska Teja Krasek were presented at many international exhibitions and published in international journals (*Leonardo Journal, Leonardo Online* and so on).

Fig. 4.25. Matjuska Teja Krasek at her exhibition *Kaleidoscopic Fragrances*, Ljubljana, 2005.

Matjuska Teja Krasek's artistic creativity is associated with various types of symmetry, Penrose tiles and rhombuses, quasicrystals, the golden section as the main symmetry element, Fibonacci numbers, etc.

With the help of reflection, imagination and intuition, she tries to find new relationships, new levels of structure, new and different types of order in these elements and structures. In her works, she makes extensive use of computer graphics as a very useful tool for creating artwork, which is a link between science, mathematics and art.

A large number of artistic compositions of Matjuska Teja Krasek is devoted to Penrose's quasicrystals and lattices (Fig. 4.26).

In Krasek's *Stars for Donald* composition (Fig. 4.27), we can observe the endless interaction of Penrose diamonds, pentagrams, pentagons, decreasing towards the center point of the composition. The golden ratio relationships are represented in many different ways in different scales.

The artistic compositions of Matjuska Teja Krasek attracted great attention of representatives of science and art. Her art is equated with the art of Maurits Escher and she is called the Slovenian "South European Escher" and "Slovenian Gift" to world art.

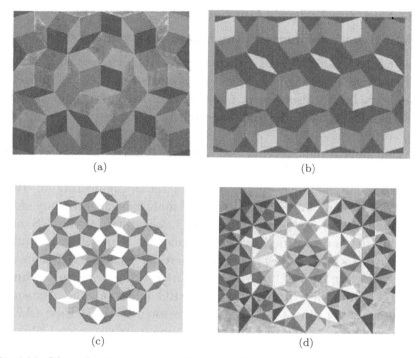

(a)                                    (b)

(c)                                    (d)

Fig. 4.26. Matjuska Teja Krasek's World: (a) The World of Quasicrystals, 1996; (b) Stars, 1998; (c) 10/5, 1998; (d) Quasicube, 1999.24.

Fig. 4.27. Matjuska Teja Krasek's *Stars for Donald*, 2005.

## 4.8. Dan Shechtman's Quasicrystals

### 4.8.1. Brief information about Dan Shechtman

The Israeli scientist Dan Shechtman is the winner of the 2011 Nobel Prize in Chemistry. He was awarded the Nobel Prize in Chemistry for his discovery of patterns in atoms called *quasicrystals*, the chemical structure that researchers previously thought was impossible (Fig. 4.28).

Dan Shechtman was born in 1941 in Tel Aviv (Palestine); this city became part of the new state of Israel in 1948. He studied at the Technion, gaining a BSc in Mechanical engineering in 1966, and MSc (1968) and PhD (1972) in materials engineering. After working in the Aerospace Research Laboratory in the US as a metallurgist, he returned to the Technion as a member of staff in the Materials Science Department (where he is currently Professor). After receiving his PhD in Materials Engineering from the Technion in 1972, Prof. Shechtman was the NRC fellow at the Aerospace Research Laboratories at Wright Patterson Air Force Base (AFB), Ohio, where he studied for 3 years the microstructure and physical metallurgy of titanium aluminides. In 1975, he joined the Department of Materials Engineering at Technion. In 1981–1983, he was in sabbatical at Johns Hopkins University, where he studied rapidly solidified aluminum transition metal alloys, in a joint program with

Fig. 4.28. Israeli scientist, Dan Shechtman.

NBS. During this study, he discovered the Icosahedral Phase, which opened the new field of quasiperiodic crystals.

### 4.8.2. What are quasicrystals?

As is known, according to the basic law of crystallography, strict limitations are imposed on the crystal structure. According to classical ideas, the crystal is composing *ad infinitum* from a *single cell*, which should tightly (face to face) "*covers*" the entire plane without any restrictions. Such were the canons of traditional crystallography, which existed before the discovery of the unusual alloy of aluminum and manganese, called *quasicrystal.*

According to the classical laws of crystallography, dense filling of the plane can be carried out by using *triangles* (Fig. 4.29(a)), *squares* (Fig. 4.29(b)) and *hexagons* (Fig. 4.29(d)). With the help of *pentagons*, such a filling is impossible (Fig. 4.29(c)).

For the first time, quasicrystals were observed by Dan Shechtman in the experiments on the diffraction of electrons on the fast-cooled Al6Mn alloy, conducted on April 8, 1982. For this discovery, Dan Shechtman was awarded the Nobel Prize in Chemistry in 2011. Such an alloy is formed during ultra-fast cooling of the alloy at a speed of 106 K per second. In this case, in the diffraction study of such alloy, the ordered pattern appears on the screen, which is characteristic for the symmetry of the *icosahedron*, which has the known forbidden fifth-order symmetry axes.

The first quasicrystalline alloy, discovered by Shechtman, was named "*shechtmanite*". The resulting diffraction pattern contained

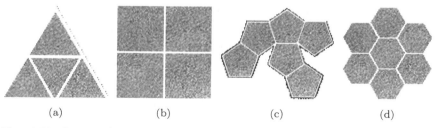

|  (a)  |  (b)  |  (c)  |  (d)  |

Fig. 4.29. Dense filling of the plane can be carried out with the help of (a) triangles, (b) squares, (c) pentagons and (d) hexagons.

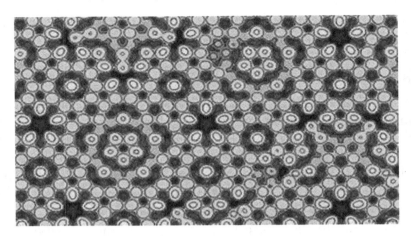

Fig. 4.30. Atomic model of a quasicrystal.

a typical pattern for crystals, but on the whole, it had a point symmetry of the *icosahedron*, that is, in particular, it had a fifth-order symmetry axis, prohibited according to the classical laws of crystallography. It was shown that the symmetry of quasicrystals is present on all scales, up to the atomic scale (Fig. 4.30), and unusual substances are indeed a *new structure for the organization of matter*.

It is important to note that the concept of a quasicrystal is of fundamental interest because it generalizes and completes the definition of the concept of a crystal. The theory, based on this concept, replaces the eternal idea about a structural unit, repeated in space in the strictly periodic manner, by the key notion of a *long-range order*.

As emphasized in the article "*Quasicrystals*" by the famous physicist D. Gratia [115], "*this concept led to the expansion of crystallography, the rediscovered riches of which we are just beginning to study. Its value in the world of minerals can be put on a par with the addition of the concept of irrational numbers to rational ones in mathematics.*"

### 4.8.3.  Experimental study of quasicrystals

During the next several years, many scientific groups around the world began studying this unusual alloy by means of high-resolution electron microscopy. All of them confirmed the ideal homogeneity of matter, in which the fifth order symmetry was preserved in macroscopic regions with sizes close to the size of atoms (several tens of nanometers).

According to modern views, the following model has been developed for obtaining the crystal structure of the *quasicrystal*. The concept of the "basic element" is the basis of this model. According to this model, the *internal icosahedrons* of aluminum atoms are surrounded by the *external icosahedrons* of manganese atoms. The icosahedra are linked by octahedra of manganese atoms.

The "basic element" has 42 aluminum atoms and 12 manganese atoms. In the process of solidification, there is a rapid formation of the "basic elements", which are quickly connecting with each other by the rigid octahedral "bridges". Recall that the faces of the *icosahedron* are equilateral triangles. To form the octahedral manganese "bridge", it is necessary that two such triangles (one in each cell) come close enough to each other and line up in parallel. As a result of this physical process, a quasicrystalline structure with the *"icosahedral" symmetry* is formed.

In the recent decades, many new types of quasicrystalline alloys have been discovered. In addition to having *"icosahedral"* symmetry (of the fifth order), there are also alloys with the *"decagonal"* symmetry (of the 10th order) and the *"dodecagonal"* symmetry (of the 12th order). Recall that the physical properties of quasicrystals began to be investigated only recently.

What is the practical significance of the discovery of quasi-crystals? As noted in the above-mentioned Gratia article [115], *"the mechanical strength of quasicrystalline alloys increases dramatically; the lack of periodicity leads to a slower spread of dislocations compared to conventional metals ... This property is of great practical importance: the use of the icosahedral phase will make it possible to obtain light and very strong alloys by introducing small particles of quasicrystals into the aluminum matrix."*

### 4.8.4. Penrose's tiles

When Dan Shechtman provided experimental evidence for the existence of the quasicrystals with the *"icosahedral" symmetry,* physicists in search of a theoretical explanation for the phenomenon of quasicrystals paid attention to the mathematical discovery made 10 years earlier by the English mathematician, Roger Penrose. As a "flat counterpart" of quasicrystals, *Penrose's tiles* were chosen, representing aperiodic regular structures, formed by "thick" and "thin" rhombuses, based on the *golden section*.

*Penrose's tiles* had been used by crystallographers to explain the phenomenon of quasicrystals. In this connection, it is necessary to note that the icosahedrons in tree-dimensional structures of quasicrystals are analogs of *Penrose's rhombuses* in plane structures.

Consider again *pentagon* and *pentagram*, which we had studied in Chapter 1 (Fig. 1.16). These geometric figures are shown in Figure 4.31.

After conducting diagonals in the *pentagon*, the original *pentagon ABCDE* can be represented as a set of three types of geometric figures. In the center, there is the new *pentagon FGHKL* formed by the intersection points of the diagonals. In addition, the *pentagon* in Fig. 4.31 includes the five *isosceles triangles* of type 1 (*AGF, BGH, CKH, DLK, EFL*) and the five *isosceles triangles* of type 2 (*ABG, BCH, CDK, DEL, EAF*).

The triangles of the first type are the "*golden*" one because the ratio of the hip to the base is equal to the *golden proportion*; they have the *acute* angles of 36° at the apex and the *acute* angles of 72° at the base. The triangles of the second type are also the "*golden*"

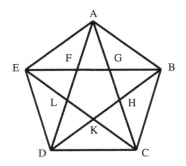

Fig. 4.31. Pentagon and pentagram.

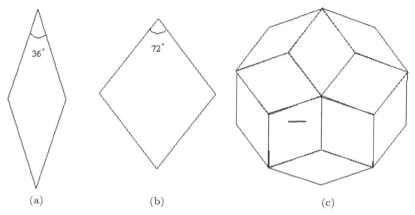

Fig. 4.32. The "golden" rhombuses: (a) "thin" rhombus; (b) "thick" rhombus; and (c) Penrose mosaic.

one because the ratio of the base to the thigh is equal to the *golden proportion*; they have the *obtuse* angle of 108° at the top and *acute* angles of 36° at the base.

Now, we connect the two isosceles triangles of the first type with their bases, and then we will do the same with the isosceles triangles of the second type. As a result, we get the two types of the *"golden"* rhombuses (Fig. 4.32). Each *"golden"* rhombus of the first type, called a *"thin"* rhombus (Fig. 4.32(a)), has the angles of 36° and 144°, the *"golden"* rhombus of the second type, called a *"thick"* rhombus (Fig. 4.32(b)), has the angles of 108° and 72°. Figure 4.32(c) shows the beginning of the construction of the Penrose mosaic.

Let's take the five "golden" rhombuses of type (b) and form a *pentagonal star* from them (Fig. 4.32(c)). After this, we add to the pentagonal star the five *"golden"* rhombuses of type (a). As a result, we get the *regular decagon*. By continuing this process, that is, by attaching the new *"golden"* rhombuses to the decagon, we can cover the plane by using only two types of rhombuses (a) and (b). In this case, there appears a certain aperiodic structure called *Penrose's mosaic* (Fig. 4.33).

The mosaic in Fig. 4.33 is named after the English mathematician and physicist Roger Penrose (Fig. 4.34), who was interested in the

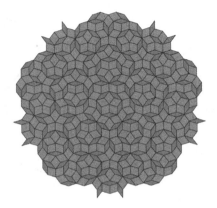

Fig. 4.33. Penrose's mosaic.

Fig. 4.34. Roger Penrose stands on the floor covered with Penrose mosaic.

problem of "tiling", that is, of filling the plane with figures of one shape without gaps and overlaps.

The classical solution of such a problem (the "parquet problem") consisted in tiling the plane with figures (Fig. 4.34), creating the recurrent mosaic. Penrose wanted to find just such figures that would not create repeating patterns, when tiling the plane. Previously, it was believed that there were no such tiles, from which the *non-periodic mosaic* could be built. Penrose's solution consisted in the following: He found only two rhombuses-shaped figures (Figs. 4.32(a) and 4.32(b)), which are based on the principle of the *golden triangle*. The resulting patterns have a quasicrystalline form, which has the fifth order axial symmetry (Fig. 4.34).

It is important to emphasize that Penrose's mosaic has the *"icosahedral"* *symmetry* or the *fifth-order symmetry*, and the ratio of the number of the *"thick"* *rhombuses* to the number of the *"thin"* *rhombuses* in Penrose's mosaic tends to the *golden ratio!*

As it turned out, Penrose's mosaic is a good analog of a quasicrystal. The three-dimensional space of the quasicrystal is filled with elementary cells just as the two-dimensional parquet is filled with the *"golden"* *rhombuses*. Penrose's idea of the dense filling of the plane with the help of the *"golden"* *rhombuses* (Figs. 4.32(a) and 4.32(b)) was transformed into three-dimensional space. At the same time, the role of the *"golden"* *rhombuses* in the new spatial structures is played by *icosahedons*. These spatial structures are called *"quasicrystals"* or *"shechtmanites"*.

### 4.8.5. Methodological significance of the discovery of quasicrystals

In conclusion, it is appropriate to make here one important remark concerning the history of the development of the *"dodecahedral-icosahedral doctrine"*. As mentioned above, the prominent German mathematician Felix Klein, back in the last century, predicted the outstanding role of the icosahedron in the development of science [113]. It is curious that this prediction was made in 1884, that is, exactly 100 years before the publication of *Shechtman's article* in the journal "Physical Review Letters" about the discovery of quasicrystals.

What is the methodological significance of this discovery? First of all, the discovery of quasicrystals is the moment of the great triumph of the *"dodecahedral-icosahedral doctrine"* [113], which permeates the entire history of natural sciences and is the source of the deep and useful scientific ideas. Second, quasicrystals destroyed the traditional idea of the insurmountable watershed between the world of minerals, in which the "pentagonal" symmetry was forbidden, and the world of living Nature, where the "pentagonal" symmetry is one of the most common. We should not forget that the main proportion of the icosahedron is the *golden proportion*, and the discovery of quasicrystals is another scientific confirmation that, perhaps, the *golden proportion*, which manifests itself both in the world of living Nature and in the world of minerals, is the main proportion of the Universe.

## 4.9. Fullerenes

### 4.9.1. What are "fullerenes"?

Now, let's focus on another outstanding modern scientific discovery in the field of chemistry [116]. This discovery was made in 1985, that is, a few years after quasicrystals. We are talking about the so-called *fullerenes*. The term "fullerenes" refers to the closed molecules of the type $C_{60}$, $C_{70}$, $C_{76}$, $C_{84}$, in which all the carbon atoms are on the spherical or spheroidal surface. In these molecules, the carbon atoms are located at the vertices of the regular hexagons or pentagons that cover the surface of the sphere or spheroid. The central place among fullerenes is occupied by the molecule $C_{60}$, which is characterized by the *greatest symmetry* and, as a consequence, the *greatest stability*.

In the molecule in Fig. 4.35(a), which has the structure of the *regular truncated icosahedron* (Fig. 4.35(a)) and is similar to the soccer ball (Fig. 4.35(b)), the carbon atoms are located on the spherical surface at the vertices of 20 regular hexagons and 12 pentagons, so that each hexagon bordered with three hexagons and three pentagons, and each pentagon bordered with hexagons.

### 4.9.2. A history of discovery

In 1985, a group of scientists, Robert Curl, Harold Kroto, Richard Smalley, and Sean O'Brien investigated the mass spectra of graphite

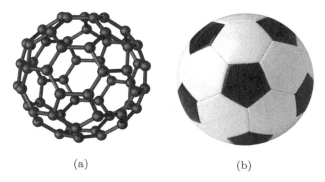

(a)                                    (b)

Fig. 4.35. The molecule $C_{60}$ and a soccer ball.

vapors, obtained by laser irradiation of the solid sample, and found peaks with a maximum amplitude, corresponding to the clusters, consisting of 60 and 70 carbon atoms. They suggested that these peaks correspond to the molecules $C_{60}$ and $C_{70}$, and hypothesized that the molecule $C_{60}$ has the shape of the *truncated icosahedron*. The polyhedral carbon clusters are called *fullerenes*. In 1966, a group of scientists (Robert Curl, Harold Kroto, Richard Smalley) were awarded with the Nobel Prize in Chemistry for the discovery of *fullerenes*.

The term "fullerene" originates from the name of the American architect, Buckminster Fuller (1895–1983), who, it turns out, used such structures when designing the domes of buildings (another use of the *truncated icosahedron!*). In 1942, Fuller developed a new cartographic presentation of the world (Fig. 4.36), composed of six rectangles and eight triangles, which had several advantages compared with the classical globe.

Since 1947, Fuller has developed a spatial structure of the *geodesic dome*, which is the *hemisphere*, assembled from *tetrahedrons*. The conception of the *geodesic dome* brought Fuller international recognition. The *golden geodesic dome* was built in 1959 for the American National Exhibition in Moscow, and in 1967 for the US pavilion at the World Exhibition in Montreal (Fig. 4.37).

Fuller taught at the Southern Illinois University from 1959 to 1970 at the School of Art and Design. In 1965, Fuller discovered the

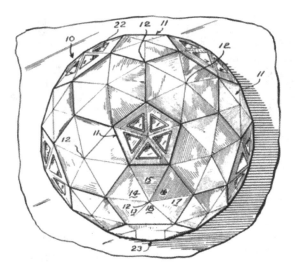

Fig. 4.36.  Cartographic picture of the world.

Fig. 4.37.  US Pavilion at the World Exhibition in Montreal (1967).

World Decade of Scientific Design (from 1965 to 1975) at the meeting of the International Union of Architects in Paris. The decade was, in his own words, devoted to applying the principles of science to solving the problems of humanity.

### 4.9.3. Fullerene applications

There are many applications of fullerenes: *optical shutters, materials for semiconductor technology,* and so on. Research in this area is being conducted in the direction of creating *superconducting materials.* Among other interesting applications, it is necessary to mention electric batteries, in which *fullerenes additives* are used, which gives positive effect.

At present, active researches on the use of fullerenes in various fields continue. Despite this, at this stage, it is still difficult to determine the most effective areas of using fullerenes.

The Russian scientists, A.V. Yeletsky and B.M. Smirnov, in the article "Fullerenes" [116], note as follows: *"Fullerenes, whose existence was established in the middle of 80s years of the 20th century, and the effective isolation technology of which was developed in 1990, are now the subject of intensive research by dozens of scientific groups. The results of these studies are closely watched by applied firms. Since this carbon modification brought a number of surprises to scientists, it would be unwise to discuss the predictions and possible consequences of studying fullerenes in the next decade, but we should be prepared for new surprises."*

## 4.10. Platonic Solids and New Ideas in the Theory of Elementary Particles

### 4.10.1. Icosahedron and elementary particles

In the recent years, the amazing symmetries of the *Platonic solids* attracted the close attention of theoretical physicists and specialists in the theory of *elementary particles* [118–120].

The prominent Russian specialist in high energy physics, academician L.B. Okun wrote as follows [120]:

> *"Physicists can be called symmetry hunters; in some sense, they differ from other people, because they are searching for more and more hidden types of symmetries in Nature."*

The main methodological significance of the article [120] consists in the fact that this article calls for theoretical physicists to use the

*Platonic solids* for the creation of the modern theory of *elementary particles*, and this fact is a reflection of the *harmonic ideas* of the ancient Greeks in theoretical physics.

The article [120] ends with the section "Forward to Plato!", where we can read the following:

> *"Among the Platonic solids, the icosahedron is the most interesting, and physicist are encountered with it, sometimes quite unexpectedly, in various branches of mathematics. This fact should serve as a heuristics, when working on the unified theory of elementary particles; possibly, the most sophisticated abstract structure is embodied in Nature. The use of the Platonic solids for the creation of the unified theory of elementary particles is the most important task of our days. As Werner Heisenberg wrote, "the development of physics looks as if, in the end, very simple laws of nature will be found, which Plato hoped to see."*
>
> *It is possible that these laws will be associated with the regular polyhedra. Even when knowledge of physical reality was still very scarce, there were thinkers (Plato, Kepler) who saw the key in these figures for understanding of physical reality. Probably, they make up that rearguard that is always ahead."*

### 4.10.2. Quark icosahedron of Yuri Vladimirov

In 2003, the International Conference *Harmony, Symmetry and the Golden Section Problems in Nature, Science and Art* was held in Vinnitsa (Ukraine) by the initiative of the author. At the plenary session of this conference, the lecture of the famous Moscow theoretical physicist, Prof. Yuri Vladimirov *Quark Icosahedron, Charges and Weinberg Angle* was heard. The lecture was published in the proceedings of the conference (Vinnitsa, 2003, Vinnitsa Agrarian University Press) [119].

The summary to the article is as follows:

> *"It is shown that the concept of generations of quarks and the values of the charges of the interactions of quarks are associated with discrete symmetries of the icosahedron, in the 12 vertices of which the left and right components of the quarks of six flavors are placed. When describing an icosahedron in cylindrical coordinates, there are three options for choosing the axis of symmetry: (1) through the midpoints of opposite edges, (2) through the midpoints of opposite faces, and (3) through opposite vertices. The first option allows you to define three generations of quarks, the second one allows to introduce the four charges, describing*

*Z-interactions of quarks and calculating the Weinberg angle, the third one allows determining quasi-electric charges and introducing the concept of quasi-spaces."*

So, there is again the *Platonic icosahedron* with reference to the theory of the *elementary particles*, and the author of the article is the scientist, quite well known in physical circles, representing the prestigious Department of Theoretical Physics of Moscow University and the author of the unique book [118].

Curiously, the book [118] ends with appendix *Weinberg Corner and the Golden Section*. The last sentence of the book [118] goes as follows:

*"Thus, it can be argued that in the theory of electroweak interactions, relationships arise that approximately coincide with the golden section, which plays an important role in various fields of science and art."*

# Conclusions

Volume I is the first part of this three-volume book *Mathematics of Harmony as a New Interdisciplinary Direction and "Golden" Paradigm of Modern Science*. The main goal of this three-volume book is to popularize the *Mathematics of Harmony* [6] as a new interdisciplinary direction of modern science. When writing Volume I, we have used the materials of Alexey Stakhov's book, *The Mathematics of Harmony. From Euclid to Contemporary Mathematics and Computer Science*, published by World Scientific in 2009 [6] and the book *The da Vinci Code and Fibonacci Series* published by Piter Publishing (St. Petersburg) in 2006 [46]. Volume I is a popular introduction to the problems of the *golden section, Fibonacci numbers, Pascal's triangle, and Platonic solids*. These mathematical objects, combined together, allow a deeper understanding of their role in the development of science.

Although we have presented mainly known data concerning the mathematical achievements of the ancient science, the medieval science, the Renaissance, and the science of the 19th and 20th centuries, the scientific results, described in Volume I, allow one to establish the uniqueness and special scientific importance of these mathematical results for the future development of modern science and mathematics. The following scientific results are the most important:

**Proclus hypothesis:** The Proclus hypothesis answers the question: What was Euclid's main goal in writing his *Elements*? The answer sounds unusual to many historians of mathematics: According to

191

the Proclus hypothesis, *the main goal of Euclid was to construct a complete theory of regular polyhedra, which are known in modern science as the Platonic solids.* To this end, Euclid, already in Book II, introduces the *"problem of dividing a segment into an extreme and mean ratio"* (Proposition II.11), known in the modern science as the *golden section.* Euclid addresses the *golden section* many times in the subsequent books, especially often in the final book (Book XIII), where Euclid used the *golden section* to construct mathematical theory of the *dodecahedron,* one of the most important Platonic solids. The *Proclus hypothesis* leads to the conclusion, which may be a surprise to many mathematicians. It turns out that the two directions of mathematical science follow from Euclid's *Elements*:

(1) **Classical Mathematics**, which did borrow from the *Elements* the axiomatic approach (*Euclidean axioms*), the *theory of numbers*, and the *theory of irrationalities*;

(2) **Mathematics of Harmony**, which is based on the geometric *"problem of dividing a segment in the extreme and mean ratio"* (the *golden section*), introduced by Euclid in Book II (Proposition II.11), and on the theory of the *Platonic solids*, described in Book XIII, that is, the concluding book of the *Elements.*

Thus, the *Proclus hypothesis* changes our views radically on the history of mathematics and its origin by starting from Euclid's *Elements* and raises the role of the ancient Greek idea of *Harmony*, based on the *golden section* and Platonic solids, in the history of mathematics.

Unfortunately, the *Proclus hypothesis* was ignored by many contemporary historians of mathematics, despite the opinion of the eminent Russian historian of mathematics D.D. Mordukhai-Boltovsky (1876–1952), who wrote in his comments to Euclid's *Elements* the following:

> *"A thorough analysis of Euclid's Elements convinced me that the construction of the regular solids, and even more the proof of the existence of the five and only five regular solids was, before Euclid, the ultimate goal of the work, from which the Elements originated."*

Fortunately, in the works of Western historians of mathematics, the *Proclus hypothesis* is discussed with increasing interest. In Refs. 117, 121, 122, the history of the *Proclus hypothesis* and its influence on the scientific work of Johannes Kepler are discussed. The *Proclus hypothesis* also influenced the work of the outstanding mathematician Felix Klein [113], and this gives us a hope that the truth someday will triumph and one of the main *strategic mistakes* in the history of mathematics will be corrected, and possibly the *Mathematics of Harmony* [6], dating back to Pythagoras, Plato and Euclid, can occupy its worthy place in the system of the modern mathematical knowledge and mathematical education.

**The "golden proportion" as a unique irrational number:** By investigating the representation of the *golden proportion* in the form of the infinite continued fractions, the Soviet mathematicians A.Ya. Khinchin [107] and N.N. Vorobyov [8] came to the conclusion that such the representation highlights the *golden proportion* among other irrational numbers because the golden proportion is the most slowly approximated by the rational fractions. This means that the *golden proportion* is a unique irrational number. The ancient mathematicians did not know about this, but their scientific intuition led them to the conclusion that the *golden section* is the main indicator of the mathematical Universal Harmony.

**Sensational books in Harmony and the golden section:** As mentioned, the 21st century is considered as the beginning of the *Era of Harmony*. It cannot be considered accidental that in the first two decades of the 21st century, the wonderful books were published in the field of *Harmony* and the *golden section*.

The first of them is Stakhov's book *The Mathematics of Harmony: From Euclid to Contemporary Mathematics and Computer Science* (World Scientific, 2009) [6].

In 2010, Prince Charles of Wales published the book *Harmony: a New Way of Looking at our World* [49]. In this book [49], the heir to the British throne shares his conviction that the most actual problems of mankind are rooted in our *disharmony* with Nature and that we can solve them only by restoring the balance with the natural order. One of the sections of the book [49] is devoted to the elaboration of the concept of the *Grammar of Harmony*, which includes the main mathematical constants, like the *golden proportion*, and the numerical sequences, like the Fibonacci numbers; they permeate the whole Nature and history of science.

Finally, the third publication is the book by the Armenian philosopher and physicist Hrant Arakelian *Mathematics and History of the Golden Section* (Moscow: Logos, 2014) [50].

**A new look at the origin of the theory of the Fibonacci numbers in modern mathematics:** In 1202, the famous Italian mathematician Leonardo from Pisa published the book, *Liber Abaci*. In this book, he introduced into the Western European mathematics the recurrent numerical sequence, in which each number is the sum of the two previous ones. This numerical sequence (the Fibonacci numbers) underlies the botanical phenomenon of *phyllotaxis* and from Fibonacci's time attracts great attention of researchers during several centuries, starting with Johannes Kepler and ending with the famous mathematicians of the 20th century, Verner Hoggatt, Nikolay Vorobyov and others. In the 19th century, the *Fibonacci numbers* were generalized by the French mathematician Lucas, who introduced into mathematics the *Lucas numbers*, another kind of the recurrent numerical sequence, which is universally recognized in modern mathematics. The works of the French mathematicians, Binet and Lucas, became the *launching pad* for the establishment of the American Fibonacci Association, organized by the initiative of the two American scientists, Verner Hoggatt and Alfred Brousseau.

The analysis of the scientific activities of Verner Hoggatt and Alfred Brousseau, and especially their photographs, presented on the Internet, as well as their statements in various journals, concerning the Fibonacci numbers, showed that the main motivating factor, which stimulated their interest in the Fibonacci numbers, was the well-known *mystery of phyllotaxis*, that is, a wide manifestation of *Fibonacci's spirals* on the surface of many botanical objects, such as pine cones, cacti, and sunflower heads.

In 1969, *TIME* magazine published Brousseau's article, titled "The Fibonacci Numbers", which was dedicated to the *Fibonacci Association.* This article contained a photo of Alfred Brousseau with the cactus in his hands. The cactus is of course one of the most characteristic examples of Fibonacci's phyllotaxis objects. The article also referred to other natural forms involving the Fibonacci numbers. For example, the Fibonacci numbers are found in the spiral formations of sunflowers, pine cones, branching patterns of trees, and leaf arrangement (phyllotaxis) on the branches of trees, etc.

Alfred Brousseau recommended to the devotees of the Fibonacci numbers as follows: *"pay attention to the search for the aesthetic satisfaction in them. There is some kind of mystical connection between these numbers and the Universe."*

Also, in Volume I, Verner Hoggatt's photo was presented, where Hoggatt holds a pine cone in his hands. The *pine cone*, of course, is another well-known example of the Fibonacci phyllotaxis objects, found in Nature. From this analysis, it may be reasonable to assume that Verner Hoggatt, like Alfred Brousseau, believed in the mystical connection between the Fibonacci numbers and the Universe. *In our opinion, this belief was the main factor and the primary motivation in Hoggatt and Brousseau's wish to study the Fibonacci numbers.*

But the Fibonacci numbers are closely related to the *golden proportion*, which in the ancient science was the main geometric object, expressing the *mathematical Harmony of the Universe*; at the same time, the Fibonacci numbers are the discrete analog of the *golden proportion.* Such a look at the theory of the Fibonacci numbers allows us to formulate the most important conclusions outlined in Volume I:

> *The Fibonacci numbers theory is actually a revival in modern science, one of the most important Pythagorean MATHEM of <u>harmonics</u>, which was lost in the Classical Mathematics, starting with Euclid's <u>Elements</u>.*

For the first time, such a point of view on the Fibonacci numbers theory was outlined in Stakhov's speech *The Golden Section and Modern Harmony Mathematics* delivered at *The Seventh International Research Conference on the Fibonacci Numbers and Their Applications* (Technical University, Graz, Austria, July 15–19, 1996). This speech has been selected for publication in the book *Applications of Fibonacci Numbers*, Volume 7, 1998 [61]. The most complete presentation of this approach is given in Stakhov's fundamental book *The Mathematics of Harmony. From Euclid to Contemporary Mathematics and Computer Science* [6].

**Nobel Prizes for the fullerenes (1996) and quasicrystals (2011) as highest recognition of the ancient "harmonic ideas" in modern science:** Quasicrystals were discovered by the Israeli physicist Dan Shechtman in 1982, and in 2011, he was awarded the Nobel Prize in Chemistry for this discovery. Fullerenes were discovered in 1985, and in 1996, a group of scientists (Robert Curl, Harold Kroto, Richard Smalley) were awarded the Nobel Prize in Chemistry for this discovery. One of the most well-known types of *quasicrystals*, called *schechtmanites*, possesses the *icosahedral symmetry*, that is, they are based on one of the most famous *Platonic solids*, the *icosahedron*, described in Euclid's *Elements*. On the other hand, one of the most well-known types of fullerenes, the molecule $C_{60}$, is based on the *truncated icosahedron*.

Both of the above-mentioned geometric shapes (the *icosahedron* and the *truncated icosahedron*) are based on the *golden section*, the great mathematical discovery of the ancient science. Thanks to quasicrystals and fullerenes, the ancient idea of the numerical harmony of the Universe entered into modern theoretical natural sciences, and these discoveries are inspiring examples for the future Nobel Laureates.

# Bibliography

## The books in the Field of the Golden Section, the Fibonacci numbers and the Mathematics of Harmony

[1] Shevelev I.Sh., *Meta-Language of Wildlife*. Moscow: Sunday (2000) (in Russian).

[2] Shevelev I.Sh., *The Principle of Proportion*. Moscow: Stroiizdat (1986) (in Russian).

[3] Shevelev I.Sh., Marutaev M.A., Shmelev I.P., *Golden Section. Three Views on the Harmony of Nature*. Moscow: Stroiizdat (1990) (in Russian).

[4] Soroko E.M., *Structural Harmony of Systems*. Minsk: Science and Technology (1984) (in Russian).

[5] Losev A., *The History of Philosophy as the School of Thought*. *Journal Communist* (1981), **11** (in Russian).

[6] Stakhov A.P., *The Mathematics of Harmony. From Euclid to Contemporary Mathematics and Computer Science*. Assisted by Scott Olsen. World Scientific (2009).

[7] Coxeter H.S.M., *Introduction to Geometry*. New York: John Wiley & Sons (1961).

[8] Vorobyov N.N., *Fibonacci Numbers*. Moscow: Science (1984) (first edition — 1961) (in Russian).

[9] Hoggatt V.E. Jr., *Fibonacci and Lucas Numbers*. Boston, MA: Houghton Mifflin (1969).

[10] Shestakov V.P., *Harmony as an Aesthetic Category*. Moscow: Science (1973) (in Russian).

[11] Vajda S., *Fibonacci & Lucas Numbers, and the Golden Section. Theory and Applications*. Ellis Harwood Limited (1989).

[12] Gardner M., *Mathematics, Magic and Mystery*. New York: Dover Publications (1952).

[13] Brousseau A., *An Introduction to Fibonacci Discovery.* San Jose, CA: Fibonacci Association, (1965).

[14] Huntley H.E., *The Divine Proportion: A Study in Mathematical Beauty.* Dover Publications (1970).

[15] Ghyka M.C., *The Geometry of Art and Life.* Dover Publications (1977).

[16] Stakhov A.P., *Introduction to Algorithmic Measurement Theory.* Moscow: Soviet Radio (1977) (in Russian).

[17] Stakhov A.P., *Algorithmic Measurement Theory.* Moscow: Knowledge (1979). (New in life, science and technology, Series of Mathematics and Cybernetics, No. 6) (in Russian).

[18] Rigny A., *The Trilogy of Mathematics.* Translated from Hungarian. Moscow: World (1980) (in Russian).

[19] Stakhov A.P., *Codes of the Golden Proportion.* Moscow: Radio and Communications (1984) (in Russian).

[20] Grzędzielski J., *Energetyczno-geometryczny kod Przyrody.* Warszawa: Warszawskie Centrum Studenckiego Ruchu Naukowego (1986) (in Polen).

[21] Garland T.H., *Fascinating Fibonacci: Mystery and Magic in Numbers.* Dale Seymour (1987).

[22] Kovalev F., *The Golden Ratio in Painting.* Kiev: High School (1989) (in Russian).

[23] *Noise-Immune Codes. Fibonacci Computer.* Moscow: Knowledge, Series of Radio Electronics and Telecommunications No. 6 (1989) (in Russian).

[24] Vasyutinsky N.A., *Golden Proportion.* Moscow: Young Guard (1990) (in Russian).

[25] Runion G.E., *The Golden Section.* Dale Seymour (1990) (in Russian).

[26] Fisher R., *Fibonacci Applications and Strategies for Traders.* New York: John Wiley & Sons (1993).

[27] Shmelev I.P., *The Phenomenon of Ancient Egypt.* Minsk: RITS (1993) (in Russian).

[28] Bodnar O.Ya., *The Golden Section and Non-Euclidean Geometry in Nature and Art.* Lvov: Sweet (1994) (in Russian).

[29] Dunlap R.A., *The Golden Ratio and Fibonacci Numbers.* World Scientific (1997).

[30] Tsvetkov V.D., *Heart, Golden Proportion and Symmetry.* Pushchino: ONTI RNTS RAU (1997) (in Russian).

[31] Korobko V.I., *The Golden Proportion and the Problems of Harmony of Systems.* Moscow: Publishing House of the Association of Building Universities of the CIS countries (1998) (in Russian).

[32]　Herz-Fischler R., *A Mathematical History of the Golden Number.*
New York: Dover Publications (1998).

[33]　de Spinadel V.W., *From the Golden Mean to Chaos.* Nueva Libreria
(1998) (second edition, Nobuko, 2004).

[34]　Gazale Midhat J., *Gnomon. From Pharaohs to Fractals.* Princeton,
NJ: Princeton University Press (1999).

[35]　Prechter R.R., *The Wave Principle of Human Social Behaviour
and the New Science of Socionomics.* Gainesville, GA: New Classics
Library (1999).

[36]　Koshy T., *Fibonacci and Lucas Numbers with Applications.*
New York: Wiley (2001).

[37]　Kappraff J., *Connections. The Geometric Bridge Between Art and
Science.* Second Edition. Singapore, World Scientific (2001).

[38]　Kappraff J., *Beyond Measure. A Guided Tour through Nature, Myth
and Number.* Singapore, World Scientific (2002).

[39]　Livio M., *The Golden Ratio: The Story of Phi, the World's Most
Astonishing Number.* New York: Broadway Books (2002).

[40]　Stakhov A.P., *New Math for Wildlife. Hyperbolic Fibonacci and
Lucas Functions.* Vinnitsa: ITI (2003) (in Russian).

[41]　Stakhov A.P., *Under the sign of the "Golden Section". Confession
of the Son of Studbat's Soljer.* Vinnitsa: ITI (2003) (in Russian).

[42]　Bodnar O.Ya., *The Golden Section and Non-Euclidean Geometry in
Science and Art.* Lviv: Ukrainian Technologies (2005) (in Ukrainian).

[43]　Petrunenko V.V., *The Golden Section of Quantum States and
Its Astronomical and Physical Manifestations.* Minsk: Law and
Economics (2005) (in Russian).

[44]　Dimitrov V., *A New Kind of Social Science. Study of Self-
Organization of Human Dynamics.* Morrisville Lulu Press (2005).

[45]　Soroko E.M., *The Golden Section, the Processes of Self-Organization
and the Evolution of Systems. Introduction to the General Theory of
System's Harmony.* Moscow: URSS, (2006) (in Russian).

[46]　Stakhov A., Sluchenkova A., Shcherbakov I., *Da Vinci Code and
Fibonacci Numbers.* St. Petersburg: Peter (2006) (in Russian).

[47]　Olsen S., *The Golden Section: Nature's Greatest Secret.* New York:
Walker Publishing Company (2006).

[48]　Petoukhov S.V., *Matrix Genetics, Algebras of Genetic Code, Noise
Immunity.* Moscow-Izhevsk: Research Center "Regular and Chaotic
Dynamics" (2008) (in Russian).

[49]　The Prince of Wales with Tony Juniper and Ian Scelly, *Harmony: A
New Way of Looking at Our World.* Harper Publisher (2010).

[50]　Arakelian H., *Mathematics and History of the Golden Section.*
Moscow: Logos (2014) (in Russian).

[51]  Stakhov A., Samuil Aranson. *The Mathematics of Harmony and Hilbert's Fourth Problem. The Way to Harmonic Hyperbolic and Spherical Worlds of Nature.* Lambert Academic Publishing, Germany (2014).

[52]  Stakhov A., Samuil Aranson. Assisted by Scott Olsen. *The "Golden" Non-Euclidean Geometry,* World Scientific (2016).

[53]  Stakhov A., *Numeral Systems with Irrational Bases for Mission-Critical Applications,* World Scientific (2017).

## The Articles in the Field of the Golden Section, the Fibonacci Numbers and the Mathematics of Harmony

[54]  Bergman G., A number system with an irrational base. *Mathematics Magazine* **31**, (1957).

[55]  Stakhov A.P., Redundant binary positional numeral systems, in *Homogenous Digital Computer and Integrated Structures,* No 2. Taganrog: Publishing House "Taganrog Radio University", (1974) (in Russian).

[56]  Stakhov A.P., An use of natural redundancy of the Fibonacci number systems for computer systems control. *Automation and Computer Systems* **6**, (1975) (in Russian).

[57]  Stakhov A.P., Principle of measurement asymmetry. *Problems of Information Transmission* **3**, (1976) (in Russian).

[58]  Stakhov A.P., Digital metrology in the Fibonacci codes and the Golden proportion codes, in *Contemporary Problems of Metrology.* Moscow: Publishing House of Moscow Machine-Building Institute (1978) (in Russian).

[59]  Stakhov A.P., The golden mean in digital technology. *Automation and Computer Systems* **1**, (1980) (in Russian).

[60]  Stakhov A.P., Algorithmic measurement theory and fundamentals of computer arithmetic. *Measurement. Control. Automation* **2**, (1988) (in Russian).

[61]  Stakhov A.P., The golden section in the measurement theory. *Computers & Mathematics with Applications* **17**(4–6), (1989).

[62]  Stakhov A.P., The golden proportion principle: perspective way of computer progress. *Visnyk Akademii Nauk Ukrainy* **1–2**, (1990) (in Ukrainian).

[63]  Stakhov A.P., The golden section and science of system harmony. *Reports of the National Academy of Sciences of Ukraine* **12**, (1991) (in Ukrainian).

[64] Stakhov A.P., Tkachenko I.S., Hyperbolic Fibonacci trigonometry. *Reports of the National Academy of Sciences of Ukraine* **208**(7), (1993) (in Russian).

[65] Stakhov A.P., Algorithmic measurement theory: a general approach to number systems and computer arithmetic. *Control Systems and Computers* **4–5**, (1994) (in Russian).

[66] Stakhov A.P., The golden section and modern harmony mathematics. *Applications of Fibonacci Numbers* **7**, (1998).

[67] Spears C.P., Bicknell-Johnson M., Asymmetric cell division: binomial identities for age analysis of mortal vs. immortal trees. *Applications of Fibonacci Numbers* **7**, (1998).

[68] Stakhov A.P., Mathematization of harmony and harmonization of mathematics, "Academy of Trinitarism", Moscow, El. No. 77-6567, Publ. 166897 (2011) (in Russian).

[69] Stakhov A.P., A generalization of the Fibonacci *Q*-matrix. *Reports of the National Academy of Sciences of Ukraine* **9**, (1999).

[70] Stakhov A., Matrix arithmetics based on fibonacci matrices. *Samara-Moskow: Computer Optics* **21**, (2001).

[71] Stakhov A., Ternary mirror-symmetrical arithmetic and its applications to digital signal processing. *Samara-Moskow: Computer Optics* **21**, (2001).

[72] Stakhov A.P., Brousentsov's ternary principle, Bergman's number system and ternary mirror-symmetrical arithmetic. *The Computer Journal* **45**(2), (2002).

[73] Radyuk M.S., The second golden section (1,465 ...) in Nature. in *Proceedings of the International Conference "Problems of Harmony, Symmetry and the Golden Section in Nature, Science and Art.* Vinnitsa State Agrarian University, **15**, (2003) (in Russian).

[74] Stakhov A.P., Generalized golden sections and a new approach to the geometric definition of a number. *Ukrainian Mathematical Journal* **56**(8), (2004) (in Russian).

[75] Stakhov A., Rozin B., On a new class of hyperbolic function. *Chaos, Solitons & Fractals* **23**(2), (2005).

[76] Stakhov A.P., The generalized principle of the golden section and its applications in mathematics, science, and engineering. *Chaos, Solitons & Fractals* **26**(2), (2005).

[77] Stakhov A., Rozin B., The golden shofar. *Chaos, Solitons & Fractals* **26**(3), (2005).

[78] Stakhov A.P., Golden section, sacred geometry and mathematics of harmony, in *Metaphysics. Century XXI. Collection of papers.* Moscow: BINOM (2006) (in Russian).

[79] Stakhov A.P., Fundamentals of a new kind of Mathematics based on the Golden Section. *Chaos, Solitons & Fractals* **27**(5), (2006).

[80] Stakhov A., Rozin B., The continuous functions for the Fibonacci and Lucas p-numbers. *Chaos, Solitons & Fractals* **28**(4), (2006).

[81] Stakhov A., Fibonacci matrices, a generalization of the "Cassini formula," and a new coding theory. *Chaos, Solitons & Fractals* **30**(1), (2006).

[82] Stakhov A.P., Gazale formulas, a new class of the hyperbolic Fibonacci and Lucas functions, and the improved method of the "golden" cryptography. Moscow: Academy of Trinitarism, No. 77-6567, publication 14098 (2006) (in Russian).

[83] Stakhov A., The "golden" matrices and a new kind of cryptography. *Chaos, Solitons & Fractals* **32**(3), (2007).

[84] Stakhov A.P., The generalized golden proportions, a new theory of real numbers, and ternary mirror-symmetrical arithmetic. *Chaos, Solitons & Fractals* **33**(2), (2007).

[85] Stakhov A.P., Three "key" problems of mathematics on the stage of its origin and new directions in the development of mathematics, theoretical physics and computer science. Moscow: Academy of Trinitarism, No. 77-6567, publication 14135, (2007) (in Russian).

[86] Stakhov A.P., The mathematics of harmony: clarifying the origins and development of mathematics. *Congressus Numerantium* **193**, (2008).

[87] Stakhov A.P., Aranson S.Ch., "Golden" Fibonacci goniometry, Fibonacci–Lorentz transformations, and Hilbert's fourth problem. *Congressus Numerantium* **193**, (2008).

[88] Stakhov A.P., Aranson S.Ch., Hyperbolic Fibonacci and Lucas functions, "golden" Fibonacci goniometry, Bodnar's Geometry, and Hilbert's fourth problem. Part I. Hyperbolic Fibonacci and Lucas functions and "golden" Fibonacci goniometry. *Applied Mathematics* **2**, (2011).

[89] Stakhov A.P., Aranson S.Ch., Hyperbolic Fibonacci and Lucas functions, "golden" Fibonacci goniometry, Bodnar's geometry, and Hilbert's fourth problem. Part II. A new geometric theory of phyllotaxis (Bodnar's geometry). *Applied Mathematics* **3**, (2011).

[90] Stakhov A.P., Aranson S.Ch., Hyperbolic Fibonacci and Lucas functions, "golden" Fibonacci goniometry, Bodnar's geometry, and Hilbert's fourth problem. Part III. An original solution of Hilbert's fourth problem. *Applied Mathematics* **4**, (2011).

[91] Stakhov A.P., Hilbert's fourth problem: Searching for harmonic hyperbolic worlds of nature. *Applied Mathematics and Physics* **1**(3), (2013).

[92]   Stakhov A., A history, the main mathematical results and applications for the mathematics of harmony. *Applied Mathematics* **5**, (2014).

[93]   Stakhov A., The mathematics of harmony. Proclus' hypothesis and new view on euclid's elements and history of mathematics starting since Euclid. *Applied Mathematics* **5**, (2014).

[94]   Stakhov A., The "golden" number theory and new properties of natural numbers. *British Journal of Mathematics & Computer Science* **6**, (2015).

[95]   Stakhov A., Proclus hypothesis. *British Journal of Mathematics & Computer Science* **6**, (2016).

[96]   Stakhov A., Aranson S., Hilbert's fourth problem as a possible candidate on the millennium problem in geometry. *British Journal of Mathematics & Computer Science* **4**, (2016).

[97]   Stakhov A., Fibonacci $p$-codes and codes of the golden $p$-proportions: new informational and arithmetical foundations of computer science and digital metrology for mission-critical applications. *British Journal of Mathematics & Computer Science* **1**, (2016).

[98]   Stakhov A., Aranson S., The fine-structure constant as the physical-mathematical millennium problem. *Physical Science International Journal* **1**, (2016).

[99]   Stakhov A., The importance of the golden number for mathematics and computer science: exploration of the bergman's system and the Stakhov's ternary mirror-symmetrical system (numeral systems with irrational bases). *British Journal of Mathematics & Computer Science* **3**, (2016).

[100]  Stakhov A., Mission-critical systems, paradox of hamming code, row hammer effect, 'Trojan Horse' of the binary system and numeral systems with irrational bases. *The Computer Journal* **61** (7), (2018).

## Other Publications

[101]  Kline M., Mathematics. *Loss of Certainty.* Translated from English. Moscow: World (1984) (in Russian).

[102]  Kolmogorov A.N., *Mathematics in its Historical Development.* Moscow: Science (1991) (in Russian).

[103]  *Harmony of Spheres. The Oxford Dictionary of Philosophy,* Oxford University Press, (1994, 1996, 2005).

[104]  The *Elements* of Euclid. Books I–VI. Translation from Greek and comments by DD Mordukhai-Boltovsky. Moscow-Leningrad (1948) (in Russian).

[105] The *Elements* of Euclid. Books VII–X. Translation from Greek and comments by DD Mordukhai-Boltovsky. Moscow-Leningrad (1949) (in Russian).

[106] The *Elements* of Euclid. Books XI–XV. Translation from Greek and comments by DD Mordukhai-Boltovsky. Moscow-Leningrad (1950) (in Russian).

[107] Khinchin A.Ya., *Chain Fractions*. Moscow: Fizmatgiz (1961) (first edition, 1935) (in Russian).

[108] Radoslav J., Pythagoras Theorem and Fibonacci numbers http://milan.milanovic.org/math/english/Pythagoras/Pythagoras.html.

[109] Korneev A.A., *Structural Secrets of the Golden Series*. Moscow: Academy of Trinitarism, El No. 77-6567, Publication 14359 (2007) (in Russian).

[110] Vilenkin N.V., *Combinatorics*. Moscow: Science (1969) (in Russian).

[111] Polya D., *Mathematical Discovery*. Translated from English. Moscow: Science (1970) (in Russian).

[112] Venninger M., *Models of Polyhedra*. Translation from English. Moscow: World (1974) (in Russian).

[113] Klein F., *Lectures on the Icosahedron and Solving Fifth-Degree Equations*. Moscow: Science (1989) (in Russian).

[114] Katz E.A., *Art and Science: About Polyhedra in General and the Truncated Icosahedron in Particular*. Moscow: Energy, **10–12**, (2002) (in Russian).

[115] Gratia D., Quasicrystals. *Uspekhi Fizicheskikh Nauk* **156**(2), (1988) (in Russian).

[116] Eletsky A.V., Smirnov B.M., Fullerenes. *Uspekhi Fizicheskikh Nauk* **163**(2), (1993) (in Russian).

[117] Kann C.H., *Pythagoras and Pythagoreans. A Brief History*. Hackett Publishing Co, Inc. (2001).

[118] Vladimirov Yu.S., *Metaphysics*. Moscow: Binom, Laboratory of Knowledge (2002) (in Russian).

[119] Vladimirov Yu.S., Quark icosahedron, charges and Weinberg angle. in *Proceedings of the International Conference Problems of Harmony, Symmetry and the Golden Section in Nature, Science and Art*, Vinnitsa (2003) (in Russian).

[120] Verkhovsky L.I., Platonic solids and elementary particles. *Chemistry and Life* **6**, (2006) (in Russian).

[121] Zhmud L., *The origin of the History of Science in Classical Antiquity*. Published by Walter de Gruyter (2006).

[122] Smorinsky C., *History of Mathematics. A Supplement.* Springer (2008).

[123] Knuth D., *The Art of Computer Programming (TAOCP) in $4^{\text{th}}$ Volumes*, Addison-Wesley, (1962, 1968, 1969, 1973, 2005).

[124] Turing A.M., The chemical basis of morphogenesis. *Philosophical Transactions of the Royal Society of London, B* **237**, (1952).

# SERIES ON KNOTS AND EVERYTHING

ISSN: 0219-9769

*Editor-in-charge:* Louis H. Kauffman *(Univ. of Illinois, Chicago)*

The Series on Knots and Everything: is a book series polarized around the theory of knots. Volume 1 in the series is Louis H Kauffman's Knots and Physics.

One purpose of this series is to continue the exploration of many of the themes indicated in Volume 1. These themes reach out beyond knot theory into physics, mathematics, logic, linguistics, philosophy, biology and practical experience. All of these outreaches have relations with knot theory when knot theory is regarded as a pivot or meeting place for apparently separate ideas. Knots act as such a pivotal place. We do not fully understand why this is so. The series represents stages in the exploration of this nexus.

Details of the titles in this series to date give a picture of the enterprise.

*Published:*

More information on this series can also be found at http://www.worldscientific.com/series/skae

Printed in the United States
By Bookmasters